Günter Nimtz and Astrid Haibel

Zero Time Space

Related Titles

Audretsch, J.

Entangled Systems

New Directions in Quantum Physics

357 pages with 77 figures and 5 tables

2007

Softcover

ISBN: 978-3-527-40684-5

Audretsch, J. (ed.)

Entangled World

The Fascination of Quantum Information and Computation

312 pages with approx. 66 figures

2006

Hardcover

ISBN: 978-3-527-40470-4

Aharonov, Y., Rohrlich, D.

Quantum Paradoxes

Quantum Theory for the Perplexed

299 pages with 72 figures

2005

Softcover

ISBN: 978-3-527-40391-2

Günter Nimtz and Astrid Haibel

Zero Time Space

How quantum tunneling broke the light speed barrier
With a foreword by Ulrich Walter

WILEY-VCH Verlag GmbH & Co. KGaA

The Authors

Prof. Günter Nimtz
II. Physikalisches Institut
Universität Köln
Germany
G.Nimtz@uni-koeln.de

Dr. Astrid Haibel
Division for Neutron and Synchrotron Scattering,
Institute for Materials Research
GKSS Research Centre
Geesthacht GmbH at German Electron
Synchrotron
Desy Hamburg, Germany
astrid.haibel@gkss.de

Original Edition

G. Nimtz und A. Haibel
Tunneleffekt - Räume ohne Zeit.
Vom Urknall zum Wurmloch
© 2004 Wiley-VCH GmbH & Co. KGaA, Weinheim

Translator

Dr. Hans-Joachim Nimtz
Offenbach, Germany

Cover

René Magritte, Carte Blanche
© VG Bild-Kunst, Bonn 2007
Foto: AKG Archiv für Kunst und Geschichte, Berlin

Library of Congress Card No.:
applied for

British Library Cataloguing-in-Publication Data
A catalogue record for this book is available from the British Library.

Bibliographic information published by the Deutsche Nationalbibliothek
Die Deutsche Nationalbibliothek lists this publication in the Deutsche Nationalbibliografie; detailed bibliographic data are available on the Internet at <http://dnb.d-nb.de>.

Typesetting Uwe Krieg, Berlin
Printing betz-druck GmbH, Darmstadt
Binding Litges & Dopf GmbH, Heppenheim

Printed in the Federal Republic of Germany
Printed on acid-free paper

ISBN 978-3-527-40735-4

Contents

Zero Time Space: How Quantum Tunneling Broke the Light Speed Barrier. G. Nimtz and A. Haibel

ISBN: 978-3-527-40735-4

Subtle is the Lord

Light faster than the speed of light? Light, which you stop and then let go again. Is that possible? Was Einstein wrong after all?

These are questions, which have kept the minds of physicists busy since, in 1994, one of the authors of this book, Professor Nimtz, chased Mozart's symphony No. 40 through a so-called tunneling barrier at 4.7 times the speed of light. The sound arrived at the other end in all its beauty. Emotions were running high when Nimtz and his colleague Enders presented their, then still rather vague, results of this so-called superluminal tunneling effect. The presentation was held at the annual spring-meeting of the "Deutsche Physikalische Gesellschaft" in Freudenstadt, Black Forest. At the time, I myself could not believe that the foundations of the Theory of Relativity could be shaken. Einstein had postulated: *"Nothing can travel faster than the speed of light."* Now we know better. We are witnesses of a rare event, of a change of a paradigm in physics. We have to accept that even Einstein's theory of relativity is not the Holy Grail of

Zero Time Space: How Quantum Tunneling Broke the Light Speed Barrier. G. Nimtz and A. Haibel

ISBN: 978-3-527-40735-4

physics. It is a macroscopic local theory, which precisely defines its limitations. Due to its macroscopic construction it will never be capable of giving us information about the smallest processes in nature on an atomic level. This can only be done by quantum mechanics. Because of its local nature we shall never be able to find out whether the universe is finite or perhaps infinite. Even a so-called flat or hyperbolical universe, which many think is *per se* infinite, could only command a global topology, which would leave the universe only finite. Nimtz's results, which come from the quantum quality of nature, force us to accept a truth beyond the theory of relativity, which we have not yet grasped.

Does this mean that any scientific theory is wrong, as it could be replaced by another tomorrow, turning everything upside down, contrary to what we have accepted as scientific knowledge? In recent years this suspicion, which at the same time laments the degeneration of knowledge, has been put forward with growing intensity, especially by the media. The German weekly "Die Zeit" wrote in August 2001:

> *"In view of the rapidly shrinking half-life period of knowledge in a world which transforms itself ever so fast, every big design of scientific theory has to face the necessity and at the same time impossibility, to regard its fate tomorrow as the intellectual fashion of yesterday."*

Here we are up against a myth, about the half-life of our knowledge, which has been bothering our society for some time. It suggests, that our knowledge is being declared invalid every five years by new knowledge. As our world is changing at a breathtaking speed, why not our knowledge at the same time? It sounds logical, therefore it might be true. What a grave error! Newton's theory of gravitation is still valid, even in the light of the theory of relativity. The periodic system of the elements has lost nothing of its validity over the centuries. Mathematical proof has been regarded since Pythagoras and Plato as eternal, metaphysical truth. It is also true that the quantity of scientific knowledge doubles about every five years, but knowledge which has been established in previous time is not rendered invalid by later discoveries. Instead it extends it towards frontiers which had not been discovered previously. Therefore Einstein's theory of relativity has extended theory rather than disproved Newton's theory. To this day the apple drops from the tree to the ground, not the other way round.

What is the relationship between the superluminal tunneling effect and Einstein's theory of relativity? There is no point in discussing, whether the superluminal effect contradicts Einstein's postulate or not. Einstein deals with free space, whereas the tunneling is not free space. Exactly at this point the evolution of physical theories reveals itself. As for the theory of light and gravitation, so has quantum mechanics extended the theory of relativity towards the tunneling process, forbidden ground in classical physics. So this non-classical expansion implies speeds beyond the speed of light.

Physicists could accept this reluctantly, if there wasn't another potential problem. It can be shown, that Einstein's postulate always preserves the principle of causality, which means keeping the universal sequence (order) of cause and effect forever. If *somewhere* in this universe first cause and then effect take place, then there is *no* place in this universe where an effect would be seen before its cause. This fundamental principle of our universe has been called the *"cement of the universe"* by the philosopher J. L Mackie. The problem now is: one can no longer guarantee causality for superluminal light waves in space. Effect can arrive ahead of cause. Man would exist before his birth!

Does the superluminal velocity, discovered by Nimtz, dissolve this "cement"? If superluminosity were to exist in space, it would indeed be the ultimate catastrophe in physics. However, as Einstein once remarked: *"Subtle is the Lord, but malicious He is not."* As the authors of this book prove in Chapter 5.5, even with the tunneling effect the cement stays with us *although* the tunneling signal is faster than light in vacuum. Not Einstein's postulate but common causality seems to be the fundamental principle of everything in nature.

But even if the Lord would leave us with what is absolutely necessary, the inner logic of the universe, at the same time he forces us to part with other ideas dear to us. Even if the sequence of cause and effect may be the same everywhere, its time distance, even time itself, can change for different observers in the universe. Time is not universal! My time *was* different during my shuttle mission in 1993. It was slower than that of any other person on earth. Therefore I have experienced 0.254 milliseconds less than those who had stayed at home. This could be measured with atomic clocks. That might sound paradoxical, but it was a fact. It is extremely difficult to understand this effect of time dilatation. The atomic clocks did not show another time, but time itself measured by atomic clocks went slower – which I myself did not notice. The result: From the biological point of view I am younger. To be honest not much, however, I am younger.

So physics is still good for a few surprises, this book is about one of them. I am convinced there will be a few more "cunning" surprises. But like Einstein I am convinced that our Lord will never turn the logic of this world upside down out of malice.

Prof. Dr. Ulrich Walter
D–2 Astronaut
Head of Space Technology
Technical University of Munich

Preface

At the beginning of the 20th century Albert Einstein developed the Theory of Relativity and revolutionized our ideas of space and time [1]. Erwin Schrödinger's, Werner Heisenberg's, Paul Dirac's and Wolfgang Pauli's quantum theory explained the wave-particle-dualism of light and also the almost unimaginable realization, that causality and determinism are become in the world of microphysics.

Since then, fascinating new effects of these laws of nature have been discovered again and again. The *tunneling effect* is one of these exciting phenomena, as is the *zero-time in the tunnel*, which is the subject of this book.

After the results of our research on the tunneling process were first published, an unexpectedly high interest in these activities erupted. Periodicals reported, invitations to public lectures followed. Television teams tried to film the tunneling effect in dark laboratories, school classes visited us to watch our experiments. Supporters and opponents frequently and extensively discussed the phenomenon of tunneling in numerous internet fora. Not least we are now in

Zero Time Space: How Quantum Tunneling Broke the Light Speed Barrier. G. Nimtz and A. Haibel

ISBN: 978-3-527-40735-4

possession of quite a few folders of letters in which we are advised of how to build a time machine and perpetuum mobiles, all sent to us by enthusiastic fans.

On the other hand there are some physicists like Moses Fayngold, who tried hard to show that quantum mechanical tunneling does not violate special relativity [2]. Fayngold presumes the experimental physicists not to know that the relevant quantities are signal and energy velocities.

With this book set in an historic frame we intend to present the tunneling effect and its consequences without formulae but easily understandable, from the first successful measuring of the finite nature of the speed of light by Ole Rømer to the acceleration of the speed of light through the tunneling mechanism. This has led to the technical use of tunneling, for instance in the fields of optoelectronics and semiconductor technology. We are going to demonstrate not only the philosophical and technical problems and the limits of the practical use of this astonishing process, but also its potential for device applications.

Our thanks for advice and support to get this book on its way, go to Prof. Ursula Haibel, Prof. Udo Kindermann, Prof. Peter Mittelstaedt, Mrs. Beate Neugebauer and Dr. Ralf-Michael Vetter. We acknowledge gratefully the accurate translation by Dr. Hans-Joachim Nimtz.

Cologne, August 2007 *Günter Nimtz* *Astrid Haibel*

1
Introduction

1.1
The Tunneling Process

There are many popular essays about time, history of the universe, teleportation or the possibility of time travel, but not much is reported about tunneling. However, the tunnel process is the basis of the origin of the universe, of the sunshine, and thus of life. It was discovered by Antoine Henri Becquerel (1852–1908), Figure 1.1, in 1886 while investigating the radioactive decay of atomic nuclei. This was then

Fig. 1.1 Physicist Antoine Henri Becquerel (1852–1908). He discovered the natural radioactivity. *© Bettmann/CORBIS*

Zero Time Space: How Quantum Tunneling Broke the Light Speed Barrier. G. Nimtz and A. Haibel

ISBN: 978-3-527-40735-4

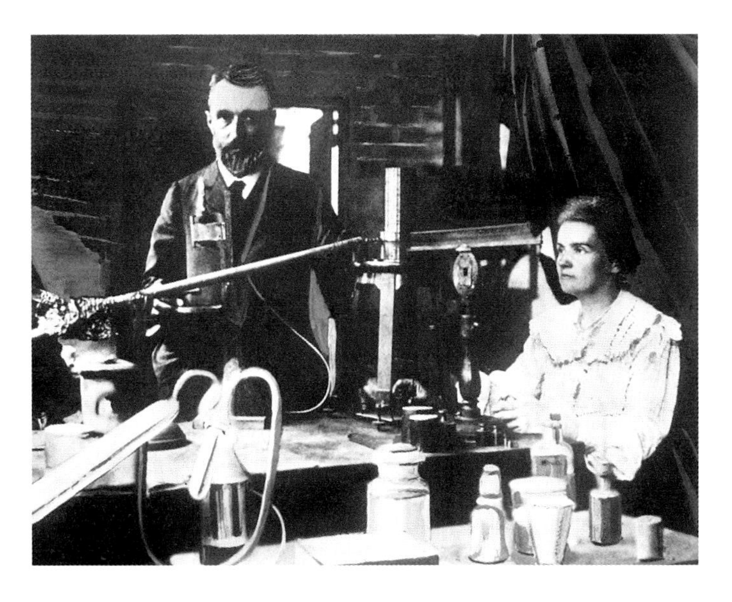

Fig. 1.2 The couple Marie and Pierre Curie (1867–1934 and 1859–1906) in the laboratory. *© Bettmann/CORBIS*

examined more systematically by Marie and Pierre Curie (1867–1934 and 1859–1906), Figure 1.2. Marie Curie was a student of Becquerel's and she continued his research. Together with Becquerel the Curies received the Nobel prize in physics *"for developing and pioneering in the field of spontaneous radioactivity and the phenomena of radiation"*. The explanation of α-decay as quantum mechanic tunneling followed around 1928 by George Gamow (1904–1975) and simultaneously, but independently, by Edward U. Condon (1902–1974) and Ronald W. Gurney (1899–1953).

Incidentally, in 1927, Friedrich Hund (1896–1997) was the first to notice the possibility of the phenomenon of tunneling, which he called *barrier penetration*, in a calculation of the ground state in a double-well potential. The phenomenon arises, for example, in the inversion transition of the ammonia molecule, as presented in Section 5.1.1.

Our knowledge today of quantum cosmology tells us that the universe also came into existence through tunneling, the so-called 'Big Bang'. A stationary state of space and time of infinitely small dimension tunneled into our world and expanded until eventually today's state of our universe was reached.

The principle of tunneling can be shown with a simple picture: Particles like photons, electrons, nuclear particles, even atoms and molecules, can surmount mountains, even though they lack the energy to reach the peak. They reach the other side of the mountain by *tunneling* through it. This process, however, is not easy to understand, the *mountains* facing those particles, have no tubes, they are not made of material that through you could get easily. Rather they are impenetrable, insurmountable barriers, which physicists call *potential barriers*. In the same way as for man a massif is insurmountable, the attraction of the particles in an atomic nucleus seem like an insurmountable barrier. These forces hold the nuclear particles together as in a potential well. For light, which consists of an ensemble of single particles of light – so-called photons – the electronic shells of atoms, for example, form a barrier. These atoms are tough obstacles for quanta of radiation, like walls for tennis balls.

Even so, after some time, particles suddenly succeed in penetrating the apparently insurmountable potential mountain. Figure 1.3 is supposed to illustrate this strange tunneling process. The thief would prefer to disappear with his loot through the wall.

Fig. 1.3 Burglars would like the tunneling effect.

This is made possible by the tunneling process, for instance during the radioactive decay of atoms. Take uranium which, emitting α and β particles (helium nuclei and electrons), decays step by step into radium and eventually into lead. This was discovered by Marie Curie. Since the 20th century radioactive decay of isotopes with a short life time has been frequently used for diagnostic and therapeutic purposes in medicine. Radioactive decay of nuclei is the source of energy in atomic power stations.

Tunneling not only makes decay possible but also enables the build up of bigger atomic nuclei, the so-called nuclear fusion. Even in the center of the sun pressures and temperatures are not high enough to bring about nuclear fusion.

The hydrogen nuclei (also called protons) cannot manage to overcome the repulsive electric barriers. Helium nuclei can not, therefore, be built up. If, in spite of this, some hydrogen nuclei do manage to reach the valley and fuse then this is also caused by tunneling. Hydrogen nuclei fuse into helium nuclei and as a result release solar energy.[1]

The most fascinating aspect of the tunneling process is that the 'tunneling' particles not only penetrate potential barriers of any size, but at the same time also show an extraordinary time behavior. Both are part of the quantum mechanical nature of the tunneling process. In this, particles seem to travel with infinitely high speed through barrier spaces, which means in zero time and thus incomparably faster than light. Light travels fast, but at a speed which is still finite and measurable. In the tunnel, however, no time exists. In a figurative sense one could say, nothing happens in a tunnel, there is eternity. Contrary to existing ideas timelessness, which is eternity, is synonymous with the total absence of events, simply boring.

Tunneling in a way is as difficult to understand as Heisenberg's relation of uncertainty. It proves that you can either measure with high precision the location or the velocity of a particle, but not both at the same time. The tunneling

1) In the most terrible weapon known to mankind, the Hydrogen bomb, both processes, nuclear decay and nuclear fusion are used. First radioactive decay is ignited. Thereby electrically positively charged atomic particles, protons, are heated to millions of degrees Celsius. Through a fusion of protons another powerful burst of energy is set free. Such a hybrid bomb can release an energy equal to 60 million tons of the classical chemical TNT (trinitrotoluene).

process and the relation of uncertainty are quantum mechanical effects which force us to revise our ideas of space and time. Since the theory of quantum mechanics was introduced about a hundred years ago no contradiction of it has been found. Of course, this theory of ours will also not be able to describe the world for ever, but ever more frequently its validity is confirmed in all fields of physics. For instance quantum mechanics is frequently used in many applications in modern semiconductor technology or in optoelectronics. Generally, standardized measures for time and signal traveling time are being used in all technologies. Therefore tunneling with its unusual time behavior provokes not only fundamental questions in theoretical and applied physics, but also philosophical and theological questions concerning space and time.

1.2 Time, Space and Velocity

In the 5th century the philosopher Aurelius Augustinus (St. Augustine), Figure 1.4, wrote about 'time' in his *confessions*:

> *"What then is time? If nobody asks me, then I know, if I want to explain it, I don't know. Even so I maintain to know confidently, there wouldn't be a past time, if nothing had passed, no future time could be, if nothing came towards us and the present time could not be experienced, if nothing existed"*. [3]

Fig. 1.4 Philosopher and Theologian St.Augustine (354–430).

This statement hits the point, time is difficult if not impossible to describe. In any case, time describes an *experience, an event*. To the classical physicist time is a *measurable experience*. (For one hour I suffered from a terrible toothache; I spent an exciting week diving in coral reefs.) Past, present and future are measurable and, for the traditional physicist, they are universally applicable, independent of place and movement.

Fig. 1.5 Portrait of Galileo Galilei around 1600. *©Bettmann/CORBIS*

Around 1600 Galileo Galilei (1564–1642), Figure 1.5, watched the oscillation behavior of a chandelier which was hanging from the ceiling in Pisa cathedral, swinging in the draught, Figure 1.6. He was curious whether the time of swinging depended on the amplitude. He did not have a watch, so he used his heartbeat and the rhythm of certain melodies for time keeping. He discovered, that the swinging time of a pendulum depended on its length, not its weight. He claimed, swinging time was independent of amplitude, which is not exactly true, as we know today.

Isaac Newton (1643–1727), Figure 1.7, assumed around 1700, like many others right up to the end of the 19th century, that time was an absolute quantity. Today we know

Fig. 1.6 Galileo Galilei studied the pendulousness of a candelabrum at the cathedral of Pisa. Painting of Luigi Sabatellio, Florenz, Museo di Fisica e Storia Naturale. *©Archivo Iconografico, S.A./CORBIS*

that this is not the case. There is no objective time, no absolute time. The passing of time depends on the movements of the observer and of what is being observed. Time was degraded by Einstein's theory of relativity to a relative quantity. In quantum mechanics time is not even an observable, which means measurable, quantity.

Although the tunnel diode has been used as an electronic device since 1962, the time which a particle spends tunneling in the mountain was neither theoretically nor experi-

Fig. 1.7 Portrait of Sir Isaac Newton in the year of 1726. *©CORBIS*

mentally revealed. In this electronic device electrons tunnel through a mountain which, in semiconductors, separates the so-called valence band from the conduction band. The mountain is called a *forbidden band gap*. The consequence of this tunneling is that this material, when in contact with a certain voltage supply, abruptly becomes a strong electric conductor. It is like using a switch, one suddenly gets a high current. The time electrons spend tunneling barriers in this often used device could not be defined until now. The cause of the problem measuring the tunneling time of electrons will be dealt with later.

Microwave signals (built up by light particles, so-called photons, which are quanta of electromagnetic waves, part of which are γ- and X-rays and light, as well as microwaves) were first measured by Achim Enders and Günter Nimtz in 1991/1992 at the University of Cologne [4]. These experiments in Cologne were provoked by an article in the journal *Applied Physics Letters* in 1991 [5]. In this, four Italian colleagues in Florence claimed that the speed of microwaves in the tunneling barrier measured by them was distinctly less than the speed of light in vacuum. Günter Nimtz, coauthor of this book, realized that these measurements could not be correct. He discussed this with Prof. Enders, at the time working with him in Cologne, who had developed a highly sensitive microwave apparatus, albeit for entirely different purposes. Enders got excited and pressed for an immediate repetition of the tunneling experiments. The following weekends were used to define the obscure tunneling speed. Initially, neither was aware of the fundamental question related to time behavior in the tunneling process. They were driven by sheer curiosity and ambition to reveal the secret of the tunneling time.

Contrary to their Italian colleagues, Enders and Nimtz observed an infinite signal velocity in the tunnel. According to their findings, the spread of the signal pulse was timeless in the barrier, it was instantaneous. The signal was spread across the entire tunneling barrier in zero time. It did not need any time to get from the entrance to the exit of the tunneling barrier. In physics this is called *non-locality* in time, in theology omnipresent. A very short time delay happens, however, at the entrance of a barrier. This time effect will be discussed in detail in Section 5.1.2.

Seen from everyday life, this surprising result collides with many respectable publications with regard to *Einstein causality*, which postulates that nothing can move faster than the velocity of light. Which means that energies and signals cannot spread faster than the velocity of light. According to most books on the special theory of relativity a velocity beyond the velocity of light would allow a manipulation of the past, see for instance Refs. [2, 6].

Einstein causality, however, can only be applied mathematically correct, to the propagation of infinitely short signals [7]. This, however, cannot be applied to particles and signals which possess a natural time duration nor to the tunneling process, which can only be explained by quantum mechanics. Not even an infinite signal velocity can change the *primitive principle of causality: cause precedes effect*. The past cannot be changed by superluminal signal velocity. (Physicists distinguish between superluminal (faster than light) and subluminal (slower than light) propagation of waves). For science fiction this means there cannot be a time machine, no manipulation of the past. In Section 5.5 the principle of causality will be discussed extensively.

By the way, the Italian scientists, who had initiated the microwave experiments as a test for tunneling velocity, confirmed the correct Cologne results in the journal *Physical Review* [8].

Quite a few physicists *believe*, that a superluminal signal velocity is impossible and *cannot be accepted*. Their comments on this subject are rather emotional. Heated disputes can be found in professional and popular science media and of course on the internet. This discussion deserves attention, as measurement of superluminal digital signal transmission was demonstrated several years ago on a modern glass fiber line in a laboratory of the *Corning* company. We shall discuss this superluminal signal experiment in more detail in Section 5.3.2.

Professor Raymond Chiao at Berkley University supports the 'impossibility' of superluminal signal transmission. He is a pupil of the Nobel prize winner Townes, who took part in the discovery of the MASER–LASER. In many articles and discussions, Chiao points out that superluminal signal transmission is simply not permissible. Interestingly, however, members of his own laboratory at Berkeley have demonstrated that quanta of light have tunneled at superluminal speeds. Measurements with the same experimental set-up confirmed that group velocity and energy velocities are superluminal [9].

At the Lake Garda meeting *"Mysteries, Puzzles and Paradoxes"* in 1998 Prof. Rudolfo Bonifacio of the University of Milan ascertained:

> *"The detector clicks faster, when a photon has gone through a tunnel than when it has traveled at the speed of light. Therefore the photon which has gone through the tunnel was registered as superluminal."*

Recently physicists like Markus Büttiker and Sean Washburn contributed to this subject in *Nature* (March 20th, 2003), claiming signal velocity must be slower than the velocity of light in a vacuum [10]. This claim cannot be correct as all light velocities are equal in vacuum, whether the velocity of energy, signal, group or phase of electromagnetic waves. In the end the authors attenuated single photons in order to rescue the subluminal signal velocity in their theoretical model. However, a single quantum of light cannot be attenuated in the elastic tunneling process. These quanta can be reflected or tunneled, but exist only virtually in a barrier as they do not spend time in traversing the barrier. In compliance with their text, the authors wanted to reduce the energy of single photons, which are the smallest units of electromagnetic waves, and thereby reduce Planck's constant?

The problem's solution lies in three often overlooked properties of the tunneling process and of a signal. We are going to deal with this in detail in Chapter 5:

- The tunneling process is part of Quantum Mechanics and cannot be described by the special theory of relativity.
- Physical signals are frequency band limited as we are familiar with, for instance, acoustic high fidelity signals having a frequency band up to 20 kHz. If the frequency band limitation were not the case, signals would need infinite energy.

- Superluminal signal speed does not violate the primitive causality saying that cause and effect cannot be exchanged.

In 1962 a publication on the theory of quantum mechanics was presented by Thomas Hartman, who described all known superluminal phenomena of tunneling and exposed all subsequent theoretical efforts as wrong or superfluous. This was dealt with at length by Steve Collins and his colleague in the *"The quantum mechanical tunneling time problem – revisited"* [11] and also by Günter Nimtz and Astrid Haibel in *"Basics of Superluminal Signals"* [12].

Recently, several physicists, first in France then in the United States, even measured negative speeds of light pulses. The result was that the peak of a pulse arrived at the exit of a medium before it had reached the entrance. Consequently, the speed of the peak traveled in the opposite direction in this special medium. Again the principle of causality does not suffer any damage, as the envelope of the pulse becomes reshaped, which means the peak of the pulse does not happen before the beginning of the original signal. Apart from that, the original information cannot be recognized anymore.

In this context it is interesting that two famous physicists, Arnold Sommerfeld (1868–1951) and Léon Brillouin (1889–1969), had already in 1914 calculated a negative group velocity in a case similar to the one just mentioned. Brillouin, however, claims in his famous book *Wave Propagation and Group Velocity* that a negative group velocity has no physical meaning [13]. Many decades later modern sophisticated electronic equipment made it possible to prove the possibility of an allegedly *non-physical* negative group velocity.

The counterpart of superluminal and negative velocities also got into the headlines recently: slowed down and stopped light. This phenomenon of slowing down light does not contradict Einstein's special relativity, but again is a quantum mechanical process which cannot be explained by the classical physics of Newton and Maxwell.

This book shows and explains the strange behavior of time and therewith velocity of tunneling. Our introductory chapters therefore deal with these fundamental physical concepts and quantities – time, space, and speed, extensively. Our understanding of these quantities has changed drastically during the last three centuries. Time is no longer regarded as absolute and signal and energy speeds are limited in their quantity by the speed of light in a vacuum. The tunnel process is something that cannot be explained by classical physics nor by the special theory of relativity, only by quantum mechanics and quantum electrodynamics.

After the introductory chapters we shall explain the spectacular qualities of tunneling. Our final sections deal with the problem of violating Einstein causality and also with superluminal phenomena, which follow speculatively from the general theory of relativity.

2
Measures of Time and Space

The change of position of a moving body over a distance in some time is generally described as *velocity*. This quantity cannot be measured directly. It derives from measuring both *time* and *distance*. We are now going to deal with how these quantities were gradually evaluated and how their measurements were fixed.

Newton (1643–1727) writes in his *"Philosophiae Naturalis Principia Mathematica"* [14] about his ideas of time and space:

> *"Absolute true and mathematical time passes in itself and according to its nature evenly and without being related to anything outside itself. It is also called duration. Relative, apparent and normal time can be felt, is external. It is either a precise or variable measurement duration. Which is generally being used like hour, day, month, year. Normal days, which generally are seen as of same duration are in fact unequal (of varying length). This disparity is corrected by astronomers, who measure movements of stars according to their real time. It is possible, that there is no constant movement, through*

Zero Time Space: How Quantum Tunneling Broke the Light Speed Barrier. G. Nimtz and A. Haibel

ISBN: 978-3-527-40735-4

which time could be measured accurately, Every movement can be accelerated or slowed down. Only absolute time cannot be changed. The same duration and the same immobility must be applied to everything existing, may they move at high or slow speed or not at all."

"Absolute space always remains the same and immovable by its nature and without any relation to an external object. Relative space is measure or movable part of the former, which through our senses and by its position towards other objects is defined and generally taken for the immovable space, that is why quite fittingly we use relative ones in human matters instead of absolute spaces and movements. In physics however one has to abstract from human senses. It could be, that there is no completely resting object, to which one could relate space and movements."

According to Newton both quantities exist without any relation to any visible object. Only towards the end of the 19th century were those comfortable and ideal theories of Newton's about space and time no longer accepted uncritically.

During the French revolution time was regarded as the great enemy. Time was the all mighty tyrant of mankind. The great German humorist and poet Wilhelm Busch describes time ironically like this: *"Is not in this world life always punished by death?"* Life is ageing. The revolutionaries wanted to destroy this *domination TIME*, too. The French revolutionaries executed many people but they failed to execute time. More modest revolutionaries usually started with a new calendar.

2.1
Measures of Time: Heartbeat, Day and Year

Time is an experience, a change of situation. We move from one place to another, we have a conversation within a certain time. Particularly important for our feeling of time is the recurring passing of a day, of a year. Time is measured by periodically passing events, day and year. As mentioned before, Galileo Galilei used his heartbeat to measure the frequency of the chandelier swinging in the cathedral of Pisa. By the way, we assume, that our time unit "second" derives from the heartbeat with its almost exact one second rhythm. From there comes the minute which is 60 seconds, the hour which is 60 minutes and lastly the day which has 24 hours. The day being the most important unit for mankind. It is interesting, that there is no mention of the origin of the second in any encyclopedia we read.

To measure time man uses periodical events like the rotation of the earth or the moon or the change of the seasons. Frequently time is measured with a pendulum: The classic upright clock uses a pendulum, which moves in the rhythm of two times a second, the same goes for a pocket watch, which uses a pendulum going to and fro, known as a "balance spring". The time unit determined by a pendulum will be a minute and then an hour and so on. Mechanical clocks achieved an accuracy of a few minutes per day. Only with these relatively accurate clocks could seamen reliably fix the longitudes on their voyages. However, they had to know the position of the sun according to the time of their home port, which was difficult in heavy seas.

Today captains at sea or in the air get these data via satellite (GPS = Global Position System). Through wireless contacts with the nearest station one can get a reliable fixing of the respective location by cellular phone. Stations to which the mobiles are connected are usually positioned at least every 30 kilometers. Time differences between the two nearest located stations allow one to locate a mobile with an accuracy of nearly one meter. [1]

Modern electronic clocks measure movements with a quartz oscillator (32 768 oscillations per second). The sum of the time of pendulum movements corresponds with seconds, minutes and hours raised to the power of the number 2. The frequency of the quartz oscillator of 32 768 Hz

1) In the 18th century the basis of exact time measuring was laid. This made England the leading seafaring nation. Until then reliable navigation was impossible due to the lack of suitable chronometers, which could determine longitudes. This urgent problem for the expanding traffic on the oceans prompted the parliament in London to grant the astronomical sum of 20 000 pounds for the development of a chronometer with an accuracy of at least 30 arc-minutes. In 1735 John Harrison developed the first sea chronometer *H1*. Since its weight (33 kg) and size (84 cm) did not meet the required specifications he only received 500 pounds. Thereafter he developed three more chronometers within the following 24 years, the last of which (*H4*) met the requirements. Its size was that of a large pocket watch (13 cm). After a voyage of five months it was only five seconds slow. Its accuracy was outstanding. He still did not get 20 000 pounds, because it was argued that this chronometer was so unique that not every ship could be fitted with one. After another five years he eventually got his 20 000 pounds. We don't know whether he was still fit enough to enjoy his fortune.

corresponds with the 15th power of 2 (2^{15} Hz), which again is a pendulum oscillation time of 0.000 030 517 578 seconds.

Atomic clocks measure extremely short and precise rotations of electrons round the nucleus of one second per 30 million years, so fast and reliable is the pendulum movement of an electron in a cesium atom. Clocks are nothing but instruments, which count periodic events over a certain time.

2.2
Measures of Length: Foot, Meter and Light Year

A measure of length was an absolute necessity in a human society, for instance to measure distances from one place to another or to measure the size of a property. Figure 2.1 shows "Foot from Frankfurt" (1575): One foot was the average length taken from the feet of *"16 respectable citizens of Frankfurt"*. More than 200 years later in 1795 the French National Assembly introduced the meter. It was the fourth millionth part of the earth's circumference along the meridian which goes through the Paris observatory. Since 1983 physicists have used a more precise fixing of a meter which is the distance light covers in the time span

$$\frac{1\ \text{meter}}{299\,792\,458\ \text{meter/second}} = 0.000\,000\,003\,336\ \text{second.}$$

Astronomers use light years when vast distances are covered: light covers about 9 460 500 000 000 000 meters or 591 281 250 000 miles in a year. The extent of our universe is

Fig. 2.1 Average length taken from the feet of *16 respectable citizens of Frankfurt*.

estimated to be about 15 billion light years, which is rather incomprehensible. In this space 125 billion galaxies have been discovered in recent years. Figure 2.2 shows a small part of the universe, in which 500 galaxies can be discovered in the shape of disks. Figure 2.3 shows a beautiful disk and spiral-shaped *whirlpool galaxy*.

A galaxy is shaped like a disk, 120 000 light years across which is 1135 000 000 000 000 000 km. The spread of a galaxy

Fig. 2.2 Picture of a part of the universe with 500 galaxies. It was produced by the overlapping of 342 pictures of the Hubble Deep Field. The bright object with a ray of beams is a star belonging to the Milky Way. To the left above this is a spiral galaxy, 2000 times further away than the Andromeda galaxy. The small star above it is one of the furthest known galaxies. *NASA/STScl*

can be disk-shaped or elliptical. One single galaxy consists of about 300 billion stars. Only since Galileo Galilei do we know of the existence of the Milky Way as a spiral and disk-shaped formation of many stars, one of which is our sun. As mentioned before, we now know that there are many

Fig. 2.3 Whirlpool Galaxy M 51. NASA and the Hubble Heritage Team. *STScl/AURA*

billions of galactic systems in the universe. Astronomers look for earth-like planets in our and other galaxies with unknown forms of life. With such an enormous number of galaxies, the existence of life on other planets can be expected. However, communication with *extraterrestrial beings*, should they exist, would be highly problematical bearing in mind the distance of many million light years between our and their galaxies.

The largest known length is the diameter of the universe. The smallest verifiable distance, the so-called "Plancksche Länge" (length discovered by the German scientist Max Planck), is theoretically smaller by 61 points (61 orders of magnitude). This length is $1.6 \cdot 10^{-35}$ meters. This value follows from the general theory of relativity. Particles with a smaller diameter cannot interact with electromagnetic waves (light). They are excluded from the outside world like a *black hole* in cosmology. It is interesting that between the smallest and the largest distance in our universe, according to the logarithmic scale, there is the biological cell, the element of life. The size is in the range of micrometers. For a long time man believed in the geocentric world of Claudius Ptolemaeus (Figure 2.4) in the second century A.D., in which the earth was the center of the universe. For the Catholic Church this was a dogma.

At the beginning of the 16th century the Polish astronomer Nicolaus Copernicus (1473–1543), Figure 2.5, postulated a heliocentric world, in which the sun and not the earth is the center of our planets. His brilliant book *"De Revolutionibus Orbium Coelestium"*, Figure 2.6, was published in 1543, when he was on his deathbed. It was dedicated to Pope Paul III. The introduction was written by Andreas Osiander (1498–1552), an intelligent but erratic Lutheran theologian. He described Copernicus' discoveries as a "hy-

Fig. 2.4 Claudius Ptolemaeus, Greek scientist, painting by Justus van Gent around 1476, Louvre, Paris. *©akg-images/ E. Lessing*

pothetical thought pattern" in order to take the sting or *provocation* out of it. Even so, in 1616 the Vatican put it on the *"Index Librum Prohibitorum"*. It took another 100 years until the world of science accepted Copernicus' ideas. The Roman Church officially accepted the heliocentric system only towards the end of the 20th century under Pope Johannes Paul II. Did Copernicus and Galilei, who confirmed the heliocentric system in 1630, have to spend that long in purgatory? So the latter arrived at the formers discovery. In their system the earth rotates around its own axis and around the sun, which causes the change between day and

Fig. 2.5 Portrait of Nicolaus Copernicus, painting around 1520.

night and the four seasons. Galilei got into trouble with the Inquisition, which could have meant being sentenced to death. He was summoned, interrogated and arrested. Under the threat of torture he revoked Copernicus' ideas. After that the life sentence was turned into lifelong house arrest. Despite many appeals from many sides this was maintained until his death.

Only 346 years later, in 1979, did Johannes Paul II reopen the case, with the result that a commission declared, in 1992, that there had been a miscarriage of justice. At last the Vatican had revoked in the case Church versus Galilei.

REVOLVTIONVM LIB. I. 21

30 anno ſuum complet circuitum. Poſt hunc Iupiter duodecennali revolutione mobilis. Deinde Mars, qui biennio circuit. Quartum in ordine annua revolutio locum obtinet, in quo terram cum orbe

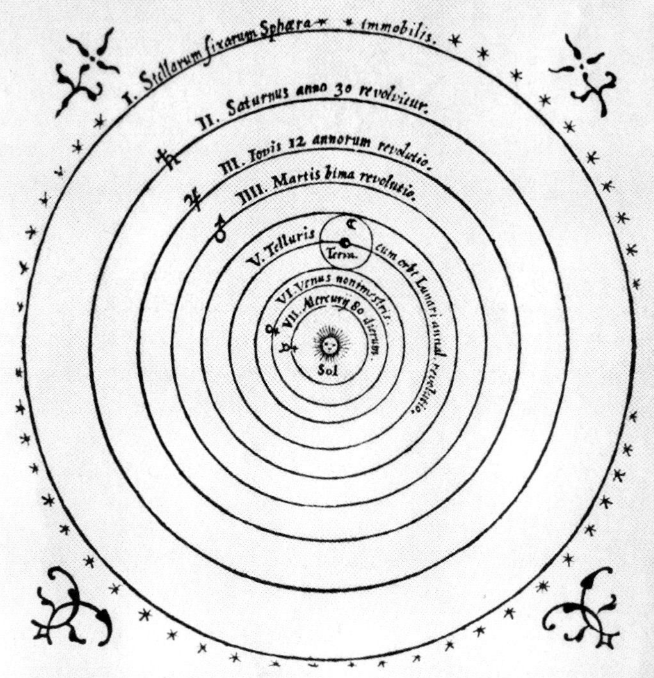

lunari tanquam epicyclo contineri diximus. Quinto loco Venus nono menſe reducitur. Sextum denique locum Mercurius tenet, octuaginta dierum ſpacio circumcurrens. In medio vero omnium reſidet Sol. Quis enim in hoc pulcherrimo templo lampadem hanc in alio vel meliori loco poneret, quam unde totum ſimul poſſit illuminare? Siquidem non inepte quidam lucernam mundi, alij mentem, alij rectorem vocant. Trimeſgiſtus *viſibilem Deum*, Sophoclis Electra *intuentem omnia*. Ita profecto tanquam in ſolio regali Sol

Solis nom[illegible] ſeu attributa.

F 5020. C 3 reſidens

Fig. 2.6 Diagram of how Copernicus perceived the world in his book *"De Revolutionibus Orbium Coelestium"* (1543). ©Bettmann/CORBIS

3
Time in Biology

3.1
Perception, Thoughts, Brainwork, Memory

Today we command gauges, created by man, which can measure time down to 10^{-15} seconds. In this extremely small space of time particle processes expire, as do processes in molecular physics. But how quickly can man, without the help of technical support, think, perceive, feel? How quickly can we react to a stimulus? In physical terms, in what minimum time can man register changes of any kind, i.e., feel? How quickly can he register light or dark, warm or cold, noisy or quiet, pleasant or unpleasant and eventually talk or melodies in a given time? Physiologists have answered these questions straightforwardly. The answer is very sobering: Man reacts and thinks lethargically and behaves in a *material way*. One could say, man reacts like sound, which is carried by matter, unlike light, which moves in an materialless vacuum a million times faster, that is 1 000 000 times faster. So light takes only 0.13 s to circle

Zero Time Space: How Quantum Tunneling Broke the Light Speed Barrier. G. Nimtz and A. Haibel

ISBN: 978-3-527-40735-4

the circumference of the earth, which is 40 000 km, whereas sound takes 1.4 days. Physicists and engineers call the slow, in relation to light, reaction of biological systems the electrical relaxation time of biology. This time shows how a disturbance of electric charge and so the electric voltage on a membrane just relaxes. Electric signals move via cell membranes and discharge stimuli of muscles or the brain. So the ear receives sound and the eye light. These will be turned into signals, which the brain can recognize and use. Nerves lead them into the brain where they are interpreted.

Physiolologists have known since the 19th century: The optimal reactions of biological systems, like nerve cells, to any kind of stimuli last about one hundredth to one tenth of a second. Thinking, feeling and reacting take place in no faster than a few milliseconds. To shorter stimuli, be they optical, acoustical, thermal, mechanical or electrical, the body does not react. It cannot feel the stimuli because our sensors can no longer cope. It is interesting that man can only perceive time when he is awake. A sleeper, even more so when under narcosis, cannot register time. In this state the human brain does not produce voltage oscillations, so called alpha- and theta-oscillations of around 5–7 Hz. These oscillations can be measured in an electro-encephalogram (EEG). This EEG registers the electrical activity of the brain and is a measure of the activities of the brain.

Higher frequencies of the brain are gamma-oscillations. Their frequency is around 40 Hz. Recently, these electrical frequencies have been allocated to the performance of the human memory. According to recent research, any perception activates several parts of the brain. First neurons (nerve cells) set off gamma-oscillations in the part of the brain, the olfactory tract, which controls smell. After that neurons transmit more gamma-oscillations in another part of the brain, the hippocampus which in turn feed the memory. If, however, the neurons of the hippocampus do not act simultaneously with the part of the brain which controls smell, then the perception does not reach the memory, which is part of the cerebral cortex. So the perception is immediately forgotten.

The highest sensitivity of a biological system lies within a frequency of around 100 Hz, the threshold of reception and of reaction to a stimulus. Incidentally, for technical and physical reasons and without knowledge of this important biological frequency, household electricity was fixed at 50 and 60 Hz, in this biologically important field.

This low frequency reflects the dynamics of human brain activity and transmission of signals in the body, for instance human reaction to internal or external stimuli. There exists, however, no evidence of any damage to the human body by household electricity, which has been in use for the last 100 years. On the contrary, man has doubled his life expectancy in this time. This proves a biological stability towards electricity with the body's natural protective devices. Otherwise the human body could be damaged by the considerable natural electric and magnetic fields in our environment.

3.2
Biological Time Unit

What we have just observed with human and other biological systems allows us to assess a biological and physical unit for everything a human being feels, sees, hears: 0.06 s. This means that, on average, man has about 40 billion perceptions during a lifetime. This quantization of a humanlife is interesting, because one would expect perceptions to occur continuously. In real life changes from joy to sadness, from light to dark happen within 0.06 s. If this took a thousand times longer, then man could watch grass growing if, however, it took a thousand times less then he could watch a bullet flying.

4
Velocity

Speed is defined as distance related to the time needed to cover it. Speed cannot be measured directly, to do this we need to measure distance and time. Driving a car at a velocity of 60 km/h we cover a distance of 60 kilometers per hour or 17 meters per second. Incidentally, at this speed a human being has a reaction time of more than two meters. From perception to reaction we drive, at this velocity, a distance of 2 meters.

What has been said about speed so far applies only to uniform motion, but in real life movement takes place at changing speeds. A car, for instance is accelerated from a standing position to a defined final velocity. Moreover the speed depends on what the traffic or road conditions allow. To ascertain the movement of an object completely all partial speeds have to be taken into account.

Modern airliners fly with a speed of about 800 km/h, the supersonic Concorde reached a maximum speed of 2 200 km/h. A pedestrian may reach 4 km/h or 2.5 miles per hour.

How fast do sound or light travel, can information be transmitted, what do the signals mean and what is their

Zero Time Space: How Quantum Tunneling Broke the Light Speed Barrier. G. Nimtz and A. Haibel

ISBN: 978-3-527-40735-4

effect? Between those two seemingly different notions, *effect* and *signal*, there is ultimately no physical difference, although they have different mathematical definitions. This can be shown by involving quantum mechanics, which will be explained in Chapter 5. In Section 4.1, definitions of different speeds will be explained.

Human voice is transmitted by sound waves. It travels with the speed of sound in the air. Through the telephone our voice is transformed into electromagnetic waves, which will reach the receiver by electric transmission, and is then converted to sound again. The advantage of this transformation of sound to electromagnetic waves is simple: electromagnetic transmission reaches its target without complications and much loss of energy over long distances by wire or wireless at a speed much faster than sound waves. As we have said before: electric waves take 0.13 seconds to get round the earth, whereas sound would take 1.4 days.

The speed of sound is tied to matter, it takes place solely in gas, liquid or in solid state matter. Like playing billiards, sound spreads by pushing one particle against another one next to it and so passes its kinetic energy. In the air it is the molecules which pass on their kinetic energy. Let us take billiards again as an example, because it illustrates the spreading of sound through molecules in the air. The central billiard ball push leaves the first ball motionless whereas the one which is hit moves on. Figure 4.1 shows with the Newton's cradle only the pulse and the kinetic energy pass on. The mass of the single spheres moves only infinitesimally out and back from the rest position. Exactly like this the

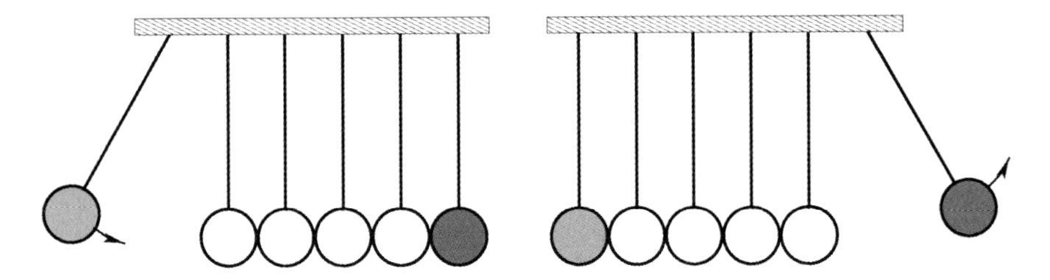

Fig. 4.1 A so-called Newton's cradle. From the left the first ball hits the first of the resting balls, but only the last one reacts by moving. All balls in between just pass on the pulse without moving. This is an example of how sound moves between closely positioned particles which do not themselves move.

spreading of sound does not transport molecules but only passes a push from one molecule to the next.

Electromagnetic waves, such as microwaves or radio waves, light, X-rays, and the high energy γ-rays, are fundamentally different from what we have described above. Electromagnetic waves do not need any medium. The difference between sound velocity and light velocity is enormous, as we have said before: Light is a million times faster than sound.

Sound velocity:	330 meter/second (in air)
Velocity of light:	300 000 000 meter/second (in vacuum)

This means that we observe lightning immediately after it hits the ground, but we only hear the thunder it caused seconds later. The shorter the time difference between lightning and thunder the nearer the spot hit by the lightning. If the lightning hits the ground some 300 meters away, we hear the thunder one second later. If the lightning hits the ground one kilometer away, we hear the thunder three seconds later.

The great Greek dramatist Aeschylus tells us, that Clytemnestra was informed of her husband's conquest of Troy the same night. We are told that the Greeks used a chain of fire to transmit news with the *speed of light*. (Thus Clytemnestra would have had ample time to prepare the murder of her husband by the time he returned to Mycenae. By the way, the interested tourist, when he visits what is left of Mycenae, is shown the bath tub in which Agamemnon was supposedly murdered.)

4.1 Velocity Definitions

Physicists and engineers distinguish between different velocities. Figure 4.2 illustrates the four most important definitions of velocities and their meaning. It shows a pulse of oscillations, which moves to the right. This packet of waves represents an electromagnetic pulse which consists of oscillations. P_1 and P_2 mark a phase and move with velocity $v_{ph} = \lambda \cdot f$, where λ is the wavelength and f the frequency of the oscillations. The *phase velocity* therefore defines the up and down of the oscillations, which form the packet.

The so-called *group velocity* v_{gr} describes, on the other hand, the speed of the spreading of the envelope of the pulse, for instance of the maximum. Group speed is thereby normally identical to signal speed v_S. While in Figure 4.2 the phase has moved from P_1 to P_2, the group has moved *"only"* from G_1 to G_2. The group has moved slightly more slowly than the phase.

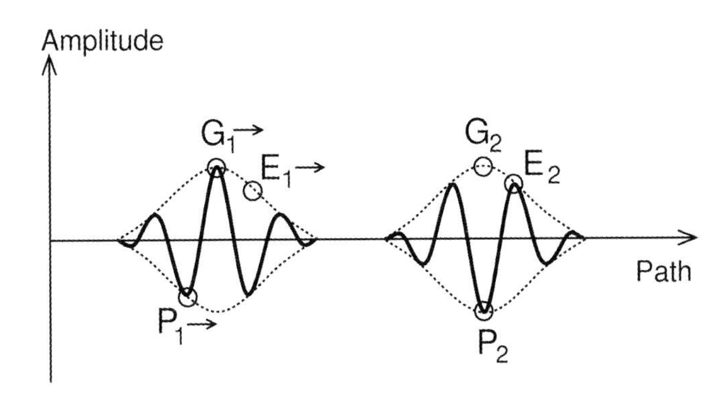

Fig. 4.2 Spreading of a wave packet. P_1–P_2, G_1–G_2 and E_1–E_2 are the distances covered in a certain time by the phase, group and signal. The latter also corresponds to the energy velocity.

Signal speed v_s is generally defined as the speed of the envelope of any wave packet. E_1 and E_2 are certain points of the signal which, for instance, convey the change in the key of a melody or its volume. In this way information is conveyed.

Not only are group- and signal-speeds generally identical, but signal speed is also the same as *energy speed* v_{en}, because all detectors, like eyes and ears, react to the energy of a signal.

Often *front speed* is mentioned as the fictitious beginning of a wave packet. However, this quantity has no physical meaning because a front does not convey information nor energy.

In a vacuum (and approximately in air) phase, group, signal, and energy velocities are equal and correspond to the velocity of light $c = 2.997\,924\,58 \cdot 10^8\,\mathrm{m/s}$ or rounded up

$c = 3 \cdot 10^8$ m/s in vacuum (air). In material like glass these four speeds are less than the speed of light. So the phase speed in a glass with refractive index $n = 1.4$ is only 71% of the speed of light. Group, signal and energy speeds are even slower, by more than 1%, than the phase speed. The ultimate extreme may be a tunnel barrier, where phase speed can be zero in the tunnel, whereas the other three velocities may become infinite.

4.2 Velocity Measurement

Generally the time is measured according to how long it takes a moving object to cover a certain distance. From distance and time speed can be deduced. With changing speed, for instance when driving a car, measuring intervals have to become shorter and shorter because the speed is changing all the time with road conditions. The speed of sound, about 300 m/s, is comparatively slow and can be measured easily. The speed of light, however, a million-fold faster, has always been controversial. Over many centuries there has been discussion on: *"How fast does light travel?"*, *"Is its speed perhaps infinite?"*.

Galilei assumed a finite speed of light, which he believed he could determine with a simple experiment. He put himself and an assistant with a lantern on the peaks of two hills, see Figure 4.3. His assistant gave a light signal and at the same time set a water clock in motion. As soon as the signal arrived at Galilei's hill, he returned the signal. As soon as Galilei's signal arrived at the other hill, the assistant stopped the water clock. This allowed them to determine the time the light had needed for the journey.

Fig. 4.3 How Galilei tried to measure the speed of light. Light moved between two hills. The time needed by the light signals should determine the speed of light.

Apart from the inaccuracy of the water clock, we know today that the reaction of the two observers was much slower than the time of light covering the distance. The result of this experiment made it clear that to define the speed of light far more precise instruments and much longer distances were needed.

The first successful fixing of the speed of light was achieved very intelligently by a young Danish astronomer, Ole Rømer (1644–1710) in 1676: Galileo Galilei had discovered four Jupiter moons in 1610, now called after the four mistresses of the ancient god: Io, Europa, Ganymed and Callisto. Galilei dedicated them to the aristocratic family of the Medici. (By the way, today we know 57 more Jupiter moons.) Figure 4.4 shows planet Jupiter with the four originally known moons.

Fig. 4.4 Jupiter and the four moons Io, Europa, Ganymed and Callisto (from above). Meanwhile 57 more Jupiter moons have been discovered. *(Courtesy NASA/JPL-Caltech)*

For navigation on the high seas in the 17th century the eclipse of the Jupiter moons was written down and the movements of those moons were thoroughly studied. To improve these investigations Ole Rømer observed, around

1675, those moons again. For this Rømer used Io, the innermost of those four moons, as a clock.

He measured the time duration of an Io day as an observer from earth. The Io day, running once around Jupiter is about 1.7 times the length of our day. He found out that when earth and Jupiter were at their shortest distance, his time tables and his measurements were very close to each other, but with growing distance between earth and Jupiter Io seemed to have a *longer day*. After half a year Io was 1000 seconds *slow passing* through the shadow of Jupiter. After another half year the clock of Jupiter was accurate again. Within a year the eclipse of Io was delayed by up to 22 minutes or approximately 1320 seconds.

It was known that Jupiter, compared to the earth, moved round the sun very slowly, a Jupiter year compares with about ten earth years. This means that within one year the earth is close and far away from Jupiter. The difference between the greatest and the smallest distance between earth and Jupiter was discovered by Johannes Kepler (1571–1630). It is 300 million kilometers. So, if someone watching from the earth gets in the course of a year away from and then nearer to Jupiter, then with a finite speed of light the eclipse of the Jupiter moons must be seen first later and then again earlier, because light will arrive first sooner and then later on the earth. So Rømer discovered, that light moves with a speed of 300 million km/1320 seconds = 230 000 km/s. Considering the technical means of that time, he got remarkably near the time we regard today as the speed of light: about 300 000 km/s. We show how it was measured in Figure 4.5.

This was an exciting result: an unimaginably fast but finite, measurable speed of light. With this result Rømer did not make many friends, because the then predominant opinion was that the velocity of light was infinite.

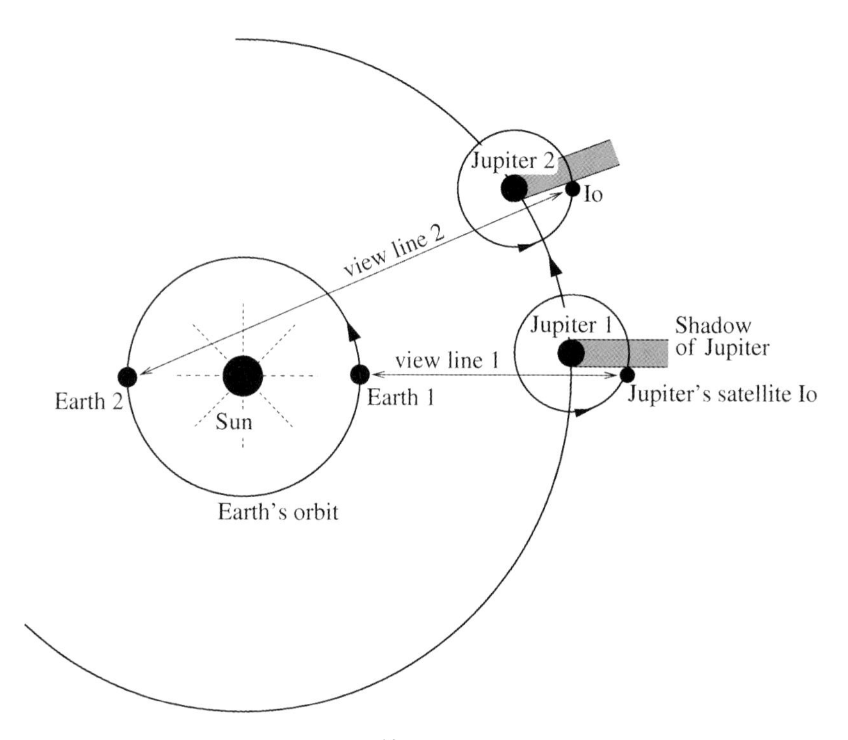

Diameter of the earth's orbit= $3 \cdot 10^{11}$ m = 300 000 000 000 m
corresponds to a measured value of 22 minutes at a velocity of

$$v_{Light} = 230\,000\,000 \text{ m/s} \quad (c = 300\,000\,000 \text{ m/s})$$

Fig. 4.5 Sketch of how the speed of light was measured with the help of the sun, earth and Jupiter with its moon by Ole Rømer in 1676. His result was 230 000 km/s, today we know it is 300 000 km/s.

4.3
Interaction Processes

The earth's force of gravitation pulls us toward it. Physicists used to describe these forces by *lines of force*, today we call them, more abstractly and generally, *lines of field*. Forces originate from physical fields. They can have these effects: Man has an effect by his weight on the scales because of the earth's gravitational pull. Electrical charged particles attract or repel each other. A force has two effects: Its strength and direction. The force or lines of field go in the direction of the interaction source.

Forces and their effects can be attributed to four different fundamental processes of interaction, which means four different origins of forces.

The process which we all know is *gravitation*. It is the force, which pulls us down to the ground, which forces the stars into their orbits. It is a general rule, mass attracts mass. As mentioned before, in the time of Newton it was generally accepted that gravitation worked *on its own* and was transferred infinitely fast. For instance all stars knew about each others existence. What could not be understood and explained was how this communication could take place in the empty space between the different stars. Today we assume, that the interaction of gravitation spreads through gravitons with the speed of light.

Gravitons are the smallest messengers of the forces of gravitation, like photons or light particles are the smallest units and messengers of electrical fields.

The finite speed of interaction in the case of electromagnetism and gravitation is contrary to the so-called instantaneous action or action at a distance, as we have learnt when dealing with sound, illustrated in Figure 4.1 "Newton's cradle".

Another well known interaction is the *electromagnetic interaction*. It is the interaction between electrically charged particles. The electromagnetic force follows the same behavior with respect to distance between the interacting particles as does gravitation, the force between the charges decreases with the square of the distance. But it has an additional quality: whereas gravitation is always attractive, the electromagnetic force can be attractive or repulsive. Because, contrary to the mass interaction of gravitation, electrical interaction can mean positive or negative electric charges. So electrons are negatively charged whereas protons, the nucleus of atoms, are positively charged. Those which are equally charged reject each other, those which are not attract each other. This interaction is often called after the French physicist Charles Augustin de Coulomb (1736–1806) *Coulomb's Interaction*. It is the basis of our life. This interaction guarantees our existence, because electromagnetic interaction is the cause of the build-up of atoms and molecules, to create liquids and solids. It is the foundation of the existence of all complex chemical materials and of all living creatures. Our body, all functions of it including thinking and dreaming, depend on or are caused by electric forces. As we have said before, it has been proved that our memory is connected with electric oscillations of our brains.

Wherever there is an electric charge moving there exist a magnetic field. An electric current, that is a moving charge, is accompanied by a magnetic field. Magnetic fields consisting of seemingly resting permanent magnets are caused by electrons which circle around an atomic nucleus. The magnetic field of the earth is due to electric currents of ions in the liquid inner part of the earth. According to this idea Magma-currents consisting of charged ions circle round the axis of the earth and cause the magnetic dipole field of the earth.

The third interaction, the so-called *strong interaction*, takes place between the components of atomic nuclei. According to *electromagnetic interaction* positively charged protons in the nucleus should repel each other because they are also positive. The atomic nucleus consists as we know of positively charged protons and electrically neutral neutrons. This repulsive effect occurs but the *total interaction* eventually becomes attractive at extremely short distances. Strong interaction therefore becomes more efficient than electromagnetic interaction at short distances. Thereby it not only overcomes the repulsive Coulomb interaction within the atomic nucleus but forces nuclear fusion. This in turn results in nuclear fusion of neutrons and protons. However, the radius of this interaction is very small so that the distance between particles which have been fused has to be very small. The diameter of an atomic nucleus is about one million times less than that of an atom. The energy of the sun emerges from the nuclear fusion process as does that of the hydrogen bomb mentioned previously. The smaller hydrogen nuclei produce the larger helium nucleus. This process releases amounts of energy incomparably larger than the chemical energy set free by burning petrol for instance. The larger helium nucleus possesses less energy than the hydrogen nuclei and neutrons put together. This fusion is possible via the tunneling process, which will be explained in Chapter 5. Even in the sun, nuclei do not have sufficient kinetic energy to overcome the electromagnetic repulsion and get within the extremely short distance of the strong interaction. The temperature at the center of the sun is 15 million degrees. To overcome the repulsive electric forces a kinetic energy of the protons is necessary which is the equivalent of a temperature of 560 million degrees. The fusion process controlled by tunneling produces a giant power which in turn is our sunlight: A sun irradiation of 1.45 kW/m^2 hits the earth.

The fourth, and to us strangest, interaction is called the *weak interaction* or weak nuclear force. It releases radioactivity and is transmitted by three so-called vector bosons. Through radioactivity large atomic nuclei like uranium or radium disintegrate into smaller ones. This process again releases energy, which is used, for instance, in atomic power stations. It is a rule that large nuclei split, releasing energy, whereas smaller ones amalgamate into larger ones. The most stable nuclei are found with atoms of the size of iron. The theory of *weak interaction* was developed by Abdus Salam and Steven Weinberg in 1967. In 1983 particles of this interaction, vector bosons, were shown experimentally. In every day life this interaction is of no importance, because it is totally overshadowed by other interactions.

The energy exchange, interaction of these four energy sources, takes place via special field quanta. In the case of gravitation they are gravitons which, however, to this day have not been proven. Proof of gravitons is being pursued internationally with great effort at present. This is very difficult, because their energy is extremely small. The energy of gravitons, the quanta of the field of gravitation, is about 40 orders of magnitude smaller than that of photons. The electromagnetic field is extremely strong compared with the field of gravitation. Forces between electric charges are incomparably stronger than those between masses. Therefore the force of attraction between two electrons because of their mass is negligible compared to the attraction or repulsion of their electric charge. The ratio of the interaction forces of masses and electric charges between two electrons is enormous: $4.17 \cdot 10^{36}$.

So in the *electromagnetic field* an energy exchange takes place via photons, with *gravitation* via gravitons, in the case of *strong interaction* via gluons and for *weak interaction* via W- and Z-bosons. Gluons and vector bosons could be proven. In every day life, apart from light quanta, these messengers of interaction do not appear.

4.4 Signals

An every day example of causing an effect or transmitting a signal is a light bulb. Here, thermally excited *glowing* atoms dispatch light quanta, which are received by the detectors of the retina and transformed into special electric signals. The brain receives and interprets them as light. This simple example of a signal tells us that the filament of a bulb is glowing, therefore hot. The reception of color corresponds to the energy of the light quanta received. Red light quanta have about half the energy of blue ones. A quantitative spectral analysis of light into its color components would also tell us the temperature of the filament. Astronomers, for instance, receive and analyze more complicated signals. From the universe they receive occasionally pulses of electromagnetic waves at all frequencies such as infrared-, X- and γ-rays, which can be seen in Figure 5.16 in Chapter 5. The frequency of radiation and the duration of the pulses give the astronomers information about a cosmic event. Frequency analyses of the pulses tell them whether they are essentially composed of infrared-, X- or γ-rays, which in turn informs the astronomers about the temperature during the event observed. (See Wien's displacement law, Section 5.1)

The duration of the radiation pulse tells us what energy was released during the event in question, whether a star collapsed or wheter two stars collided.

A modern technical signal is illustrated in Figure 4.6. It shows an amplitude modulated signal of the infrared carrier wave with a frequency of $2 \cdot 10^{14}$ Hz. (Amplitude modulation is known to us as AM from radio. Historically this technique became popular with the use of Morse code in the middle of the 19th century.) An electromagnetic wave of constant frequency, in this case an infrared wave, is modulated by a time dependent change in the amplitude. The duration of the amplitude of the carrier wave delivers the information, which is being transmitted, *carried* by that wave with a defined frequency. These signals in Figure 4.6 transmit the digits of our telephone calls and other digital data overseas, and link computers.

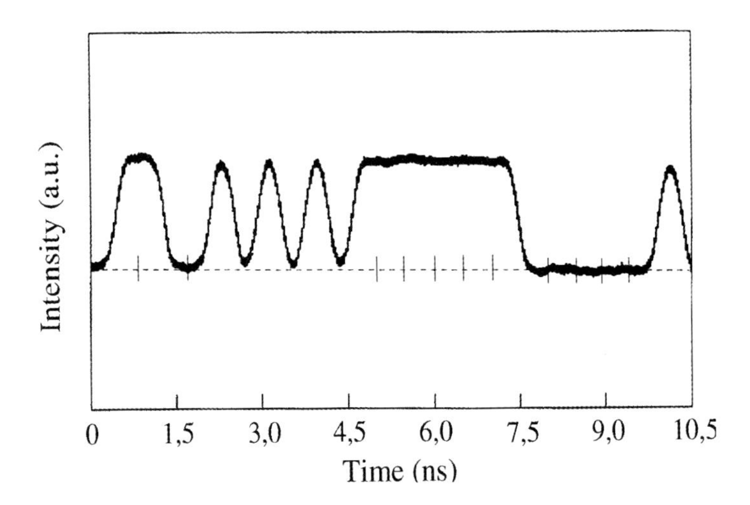

Fig. 4.6 The time dependence of the electric intensity of an AM signal transmitted through a glass fiber. A sequence of digital signals, pulses of different length, consisting of single bits can be seen.

The half width, which is the time interval at half height of the intensity of the signal, is like the good old Morse information, while the carrier frequency finds the addressed receiver. The half width, and thus the content of a signal, is independent of the absolute height of the signal or its intensity, otherwise strong signals from stations nearby would transmit different music than those weak signals from stations far away.

Today signals are not like Morse codes but mathematically more logical. Digitally coded 0 and 1, expressed through the half duration of a pulse. As Figure 4.6 shows, the signal consists of bits 1,1,0,0,1,0,1,0,1,1,1,1,1,1,1,0,0,0, etc. The smallest unit of information is called a bit. One byte consists of 8 bits. The sequence of 0 and 1 determines a number or a letter. Our language and any kind of information is translated into this digital form, transmitted and then re-transformed into understandable language. The generally used binary code uses only 0 and 1 as its base. Our decimal system has the base numbers 0 to 9. The connection between the two is shown in Table 4.1.

Tab. 4.1 Connection between binary and decimal systems.

Decimal	0	1	2	3	4	5	6	7	8	9
Binary	0	1	10	11	100	101	110	111	1000	1001

For the computer the binary system is more advantageous than the decimal one. According to their use there are other systems such as the octal-, hexa- and the duo-decimal systems. The latter consists of twelve numbers. Twelve can be divided by 2, 3, 4, and 6, ten only by 2 and 5. It is assumed that some early civilizations used the more advantageous duodecimal system. The reason why our decimal system was established is very likely because of our ten fingers.

An effect or a signal can only be transmitted or released by energy quanta (gravitons, photons, gluons or vector bosons) of the four fundamental interaction processes we have mentioned before. Detectors, measuring instruments like eyes and ears react only to quanta, which means: energy particles of light and sound with corresponding interaction. This principle applies to all fields.

4.5
From Galilei via Newton and Einstein to Quantum Physics

At the time of Galilei it was assumed, that velocities between moving transmitters and receivers would be additive. For instance, an observer would measure the velocity of the receiver plus the velocity of light. A bird flying against a window experiences less damage than a bird which hits the windscreen of a moving car – or lets put it like this: both are probably killed, but the latter has no chance at all. The speeds of bird and car are added together. According to this classic theory a bullet shot from a moving train reaches a speed which is its normal speed plus that of the train. This *addition of individual speeds*, the so-called Galilei Transformation, is valid only for *low* speeds, as was shown by Einstein's Special Theory of Relativity 300 years later. "Low" in this case means low in relation to the speed of light of 300 000 km/s or 675 000 miles per hour.

At the end of the 19th century Hendrik Antoon Lorentz (1853–1928), Figure 4.7, and others thought that this addition of two speeds does not include the speed of light as it does when a bullet is fired from a moving train. Light for the observer in a moving train travels as fast as for an observer who is stationary. That discovery was exciting and sensational. The speed of light is obviously con-

Fig. 4.7 Physicist Hendrik Antoon Lorentz (1853–1928). He introduced the Lorentz-Transformation, fundamentally important for the Theory of Relativity. Lorentz describes the ratio between resting and moving systems. This confirms the principle of the constancy of the speed of light. Together with P. Zeeman he received 1902 the Nobel prize. *©Bettmann/CORBIS*

stant, independent of transmitter or observer. That contradicted the classical Galilei-Transformation of the addition of speeds. Eminent scientists like Hendrik Antoon Lorentz and Henri Poincaré dealt with this phenomenon. Max Born commented:

> *Lorentz introduced in 1892 a "surprising hypothesis", that every body contracts in the direction of its movement (Lorentz contraction), whereas time experiences a dilatation* [1].

Eventually, in 1905 Einstein described this phenomenon mathematically in the *Spezielle Relativitätstheorie*, the special theory of relativity (STR). Between systems moving at constant but different speeds, called "inertial systems", no matter how fast they move, there will always be measured, for any observer, a constant speed of light. However, the *length* of the object as well as the *duration* of time in the moving systems change for the observer. Physicists call a system in which all laws of classical mechanics are valid an inertial system.

If an observer is stationary relative to the moving object there is a contraction of length and a dilatation of time. This observer sees in the moving system objects that appear shorter in length and slower moving than those seen by the moving observer.

With this knowledge follows, from the special theory of relativity, the unexpected phenomenon of the so-called "paradox of the twins".

This can be illustrated by a theoretical experiment: one twin is sent into space at nearly the speed of light, the other one stays on the earth. The brother traveling at high speed ages slower than the one staying on earth. However, although the traveling twin is going to live longer, he has not gained a lengthening of his life. He just experiences the approximately 40 billion sensations of a lifetime more slowly than does his brother who stayed on earth. The sum of his sensations is not changed by this journey.

Joseph C. Hafele at the Washington-University of St. Louis together with the astronomer Richard Keating at the US Naval Observatory in 1971 succeeded, by means of a simple experiment, in confirming the twin paradox, the dilatation of light. They traveled in a jet plane round the world, first going west, then going east, with four atomic clocks on

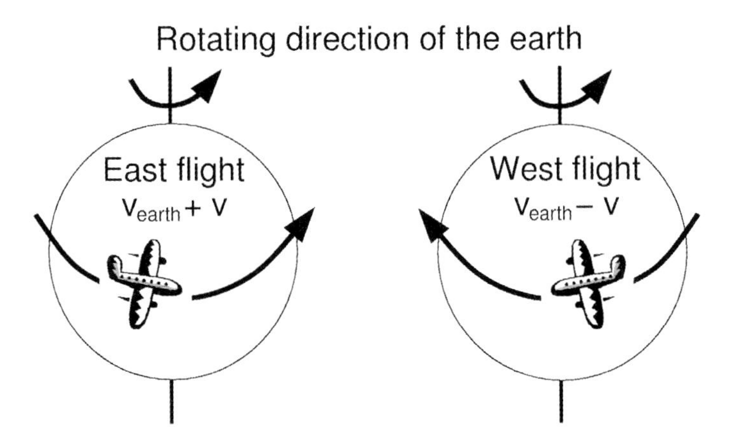

Fig. 4.8 The Hafele–Keating Experiment proved the twin paradox by traveling with atomic clocks on scheduled planes round the earth going both westwards and eastwards.

board. On the ground they left clocks to compare the time. As is shown in Figure 4.8, when flying eastwards the speed of the plane and the speed of the rotation of the earth are added. The speed of the plane appears faster to the observer than the clocks on the ground show. So time on board passes more slowly. Flying westwards the plane is moving against the rotation of the earth. Therefore the speed of the plane is subtracted from the speed of the earth. For the observer on the ground the plane is moving more slowly than the clocks on the ground, time on the plane passes more quickly. When the clocks were compared back on the ground, those clocks going west were fast, those going east were slow compared with those on the ground.

With this experiment another effect of relativity was found, which Einstein described in the "Allgemeine Relativitätstheorie", the General Theory of Relativity.[1] The clocks on board the planes should be fast simply on account of the *differences in the gravitation* in the air and on the ground. Because in the easterly direction the effect of gravitation is working against the effect of speed, the time difference should be considerably less than when going west, where one could expect a clearly larger difference. Exactly this was shown by the measured times (see Table 4.2).

Tab. 4.2 Measured and calculated time differences going round the earth in westerly and easterly directions, comparing them with clocks on the ground. Data were taken from [15].

	Measured time value Δt [ns]	Calculated time value Δt [ns]
East flight	-59 ± 10	-40 ± 23
West flight	$+273 \pm 7$	$+275 \pm 21$

The results of these flights, also taking into account the lower strength of the earth's gravitation because of the flight altitude, confirm the forecasts of the general theory of relativity.

1) The general theory of relativity goes much further than the special theory of relativity and describes the results not only in a stationary state or at constant speed, but also with accelerated systems. Special importance is given to acceleration through gravitation forces, that means the interaction between masses. Einstein did show in this theory that energy and mass are equal. He predicted that light from a far star passing our sun is bent on the way to an observer on earth due to the gravitational force of the sun. The predicted light deflection due to the sun mass was measured by Eddington in 1919.

5
Faster than Light and Zero Time Phenomena

In 1992 an exciting observation in the world of physics took place, which contradicts our normal life experience. In that year photonic experiments to define the behavior of time in the *tunnel effect* delivered strange results, such as that tunneling particles went through extensive barriers apparently *without losing time*, i.e. in zero time, although they did not have sufficient energy to overcome those obstacles. Normal experience tells us that these particles should not have been able to overcome these obstacles in the first place because they were too high for them. Why then did they get to the other side and moreover in zero time within the mountain? Figure 5.1 illustrates the tunnel effect. According to quantum mechanics it is highly unlikely that you will find *"the ball"* on the other side of the wall but there is a finite probability of finding one of the many balls on the other side.

Tunneling particles have been a well known quantum mechanical phenomenon for a long time, although rarely observed in everyday life. The timelessness of the tunnel process, however, is hardly imaginable and had never been measured before. It is remarkable that the results of accurate

Zero Time Space: How Quantum Tunneling Broke the Light Speed Barrier. G. Nimtz and A. Haibel

ISBN: 978-3-527-40735-4

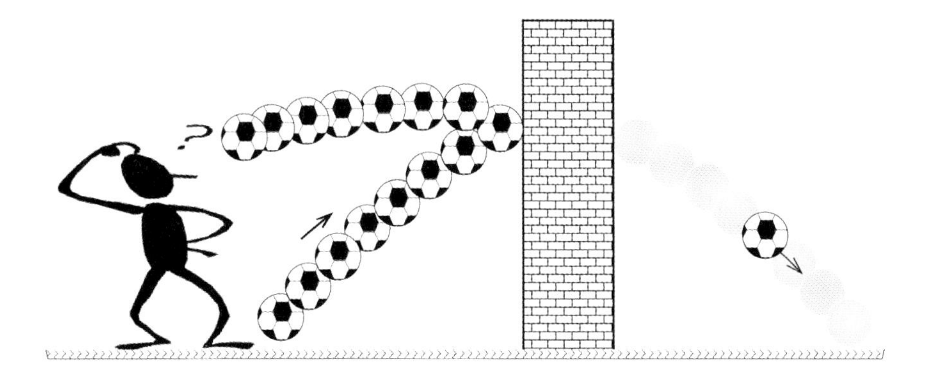

Fig. 5.1 Cartoon of the tunnel effect: how does one *"ball"* of eight get through the wall?

analysis do not contradict the theory of quantum mechanics, a theory today generally accepted. Nor do they collide with the rules of causality. Even with the tunnel effect you cannot exchange cause and effect. We shall now explain the so-called *tunnel effect*.

5.1 The Tunneling Process: Space with Zero Time

5.1.1 The Tunneling Effect

At the beginning of the 19th century it became possible to explain the tunnel effect through a new theory: the theory of quantum mechanics. It is now the foundation of modern physics.

By chance the up and coming lighting industry with the development of the light bulb helped this revolutionary research – a little. The newly developed bulb had started to replace the gas lantern. At the Physikalisch-Technische Reichsanstalt in Berlin work took place to examine the radiation qualities of the bulb. The scientists wanted to know how and why the light quantity and the color of a hot body change with rising temperatures. The higher the temperature of the radiator the more the color of the light changed from red to white.

Wilhelm K. W. Wien of the Reichsanstalt (1864–1928) succeeded in 1896 by sheer observation in discovering the law of spectral energy distribution, the color composition of radiation depending on the temperature. This law is called *Wien's displacement law*.

Max Planck (1858–1947), Figure 5.2, after Wien's empirical discovery, deduced theoretically, in 1899, the behavior of a *thermal radiator*. Since precision measurements in the region of long waves resulted in considerable divergences from *Wien's and Planck's discoveries*, Planck looked into their formula again.

He presented, first on May 8th 1899 to the "Königlich Preussische Akademie der Wissenschaften" and then again, one year later, on December 14th 1900, to the "Physikalische Gesellschaft in Berlin" a theoretically founded formula, which now defined the data of their measurements completely. For this he had to abandon the previously accepted conviction that energy was a quite arbitrarily divisible quantity. In his formula radiation energy E related

Fig. 5.2 Max Planck discovered “Qantenmechanik”. *Photo: Deutsches Museum München*

to frequency f accepted *discrete* quantities ($E = h \cdot f$). For this it was necessary to introduce a "natural constant h" $h = 6.6 \cdot 10^{-34}$ J s. This number has since become known as the *Planck constant*.

Planck, however, was not happy with his formula, because it contradicted completely the conviction of classical physics, which claimed that nature does not move in *"leaps"*. Later he wrote that he regarded all this as an *"act of despair"*.

Neither Planck nor his colleagues realized at first the enormous consequences of the *discrete nature*, of the quanta of energy. Only five years later Albert Einstein (1879–1955), Figure 5.3, presented his much more radical hypothesis of

Fig. 5.3 Albert Einstein,1912. He expanded the quantum theory of Planck by the Hypothesis of light quanta. *Photo: Deutsches Museum München*

light quanta, which meant the final break with traditional doctrines. In his "Über einen die Erzeugung und Verwandlung von Licht betreffenden heuristischen Gesichtspunkt" (dealing with a point of view concerning creation and transformation of light from a heuristic angle) he regards light as *"bombardment with particles"*. He used this theory successfully to explain *the photoelectric effect*. Einstein proved that a photon (light quantum) absorbed from an atom passes its energy on to only one electron. As shown in Figure 5.4 a photon with its energy quantum $h \cdot f$ separates one electron from an atom. The energy of the photon is passed on to the electron completely. In 1921 Einstein received for his

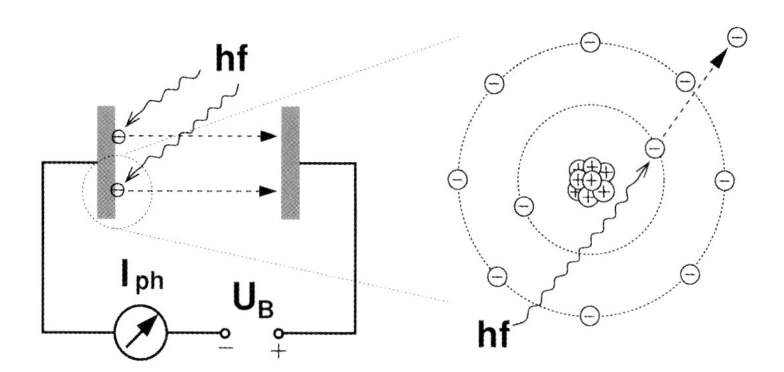

Fig. 5.4 The photoelectric effect. The photon meeting with an electron passes its energy in the form of discrete energy quanta hf. If the energy is strong enough, the electron is removed from the metal surface, resulting in an electric current.

explanation of the photoelectric quantum effect (not for his famous Theories of Relativity) the Nobel prize.

In this context it is interesting that Newton, 300 years earlier, had described light as particles, not as waves. With this he had in his time hardly any followers, since all experiments then showed light as a wave.

In 1911 at the first Solvay conference in Greece, a *summit conference* of the then leading physicists, Figure 5.5, Planck's quantum hypothesis had its breakthrough. It became the center of physics of that time.

Many effects in optics, electronics or in modern nanotechnology, and also chemical or molecular biological processes can only be explained through the quantum theory. The same goes for the functioning of the Laser, the modern light source (**L**ight **A**mplification by **S**timulated **E**mission

Fig. 5.5 First Solvay conference in 1911. At this conference Planck's quantum hypothesis had its breakthrough. *© Hulton-Deutsch Collection/CORBIS*

of **R**adiation), the reason why metal is a better conductor than other solid bodies and why atoms are stable in the first place, something that could not be explained until the arrival of the quantum theory. Why do electrons not collapse with the positively charged nuclei?

Electromagnetism, one of the four fundamental processes of interaction, was described brilliantly by James Clark

Maxwell (1831–1879) in 1873. One incompleteness was discovered only when the hydrogen atom was described. The attraction of a negatively charged electron and a positively charged proton form a hydrogen atom. Maxwell's theory does not explain why this does not lead to emission of electromagnetic waves. Instead the electron moves like a satellite round the proton, the nucleus of the atom, and is not drawn to the proton within the classically expected $3 \cdot 10^{-13}$ s. The stability of the orbit and the property of electrons to move round the nucleus in certain orbits can only be explained by quantum theory.

Important differences between quantum theory and classical physics are:

- Quantization.

  All physical fields are quantized as multiples of minimum energy quantities, so-called *quanta*. Fields can only have an effect through the quanta and only quanta are measurable.

- Dualism of waves and particles.

  Is an electron like a particle such as a billiard ball or is it like a wave? Field quanta, like electrons and photons, which have been examined, have particle as well as wave properties. Wave and particle are an inseparable unit. Light quanta, i.e. photons, can, because of their particle qualities, when hitting a metal pass on their energy to electrons and thereby separate them from the metal atom. They can, however, because of their wave qualities, moving out of a slit produce a pattern of bright and dark bands. Which phenomenon, particle or wave, is observed depends on the individual experiment – which might sound surprising.

- The end of the unequivocal predictability of an event, which means the end of determinism.

  Classical physics allows us to determine the behavior of a particle in advance. A particle possesses a certain energy and moves in a certain direction at a time. This fixes its behavior forever. With quantum physics this is no longer possible. Its future behavior is coincidental and can be predicted only with a certain probability. Quantum mechanics predicts for instance that out of a 1000 atoms about 10 will decay. We cannot, however, predict which of the 1000 radioactive atoms will decay at which time.

- The impossibility of an exact measurement (Relation of uncertainty).

  It is impossible, as one can in classical physics, to determine space and momentum or energy and time of a quantum simultaneously and exactly. The more precisely one determines one of these quantities, the less precise becomes the other one.

  This uncertainty can be explained clearly: the more precisely one wants to determine the locality of a particle the smaller the wavelength of the detecting light has to be. However, the energy of the light quantum increases in reciprocal proportion to the wavelength and interferes, while measuring, with the energy of the investigated object.

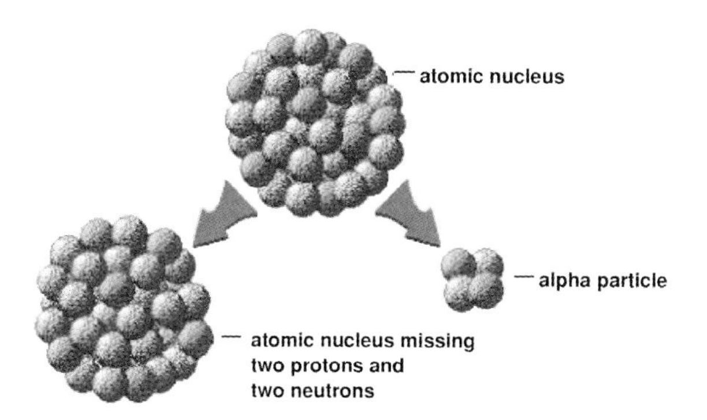

Fig. 5.6 Nuclear decay in a uranium atom. An α-particle leaves the nucleus although its energy is too small to overcome the attractive nuclear force of the nucleus.

The uncertainty between energy and time is sometimes seen as a cause of the tunnel effect. It makes it theoretically possible to give a particle, for an extremely short time interval, added energy, which allows it to overcome barriers which had not been possible before. This, however, does not explain the time behavior in the tunneling process, as will be seen later.

Let us explain the tunnel phenomenon with some examples. The tunnel effect was first discovered in 1926 with the radioactive decay of uranium and two years later was explained through quantum mechanics.

Large atom nuclei with many components are unstable and decay in the course of time into smaller ones. Giant atoms, as existed after the "big bang", decayed long ago, they cannot be found anymore. Their span of life, their time of decay, was shorter than the present age of the earth.

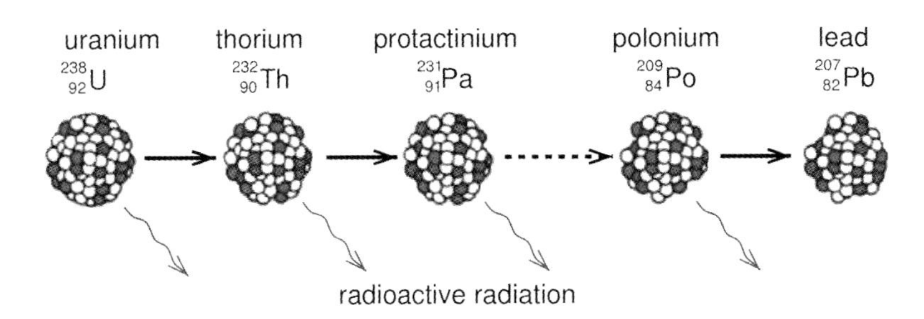

Fig. 5.7 Decay of a radioactive uranium nucleus into lead, with a much longer lifetime.

One of the most stable elements is the iron atom. It consists of 82 components, 26 protons, 56 neutrons and 26 electrons in the atom shell of quantized orbits. In this arrangement of protons that electrically repel each other and shielding neutral neutrons the energetically favorable state for the atom nucleus is reached. All natural radioactive processes, on enormous time scales, lead to larger elements decaying into iron atoms or smaller atoms with fewer elements fusing in the direction of iron atoms, for instance hydrogen to helium. During the radioactive alpha-decay a nucleus element separates from the uranium nucleus, the so-called α-particle consisting of two positive protons and two neutrons, although nuclear force should keep it inside the nucleus of the atom (see Figure 5.6).

This results in two new particles, the escaping α-particle and the remaining thorium atom. Since thorium is still radioactive, the nuclear decay continues. It continues to radiate, as shown in Figure 5.7, electrons, γ-quanta, protons and other particles and decays until the first non-radioactive, stable – with a long lifespan – element emerges: lead.

Figure 5.8 shows a very simplified picture of the tunnel process during the decay of a uranium nucleus. An α-particle presented as a wave packet moves to the other side of the mountain although its energy is too small to get over the mountain. In this case the mountains are representative of the attractive nuclear forces. The strong nuclear forces are effective only at very short distances, much smaller than an atom's diameter.

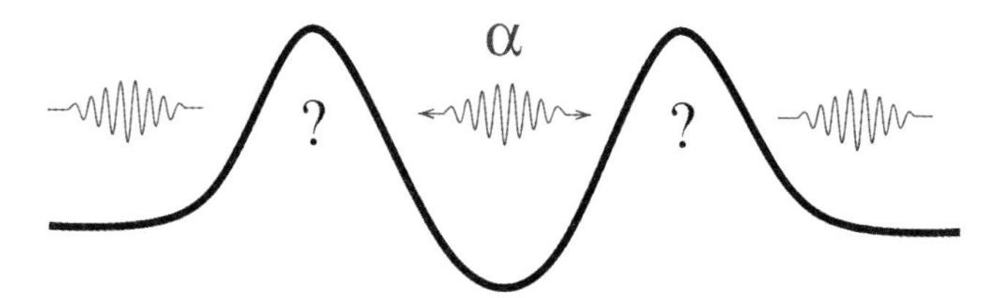

Fig. 5.8 A sketch of the tunnel effect. In the valley confined by two mountains an α-particle is arrested. Although having too small an energy to escape sometimes the particle is found outside the mountains: physicists call the process tunneling. We are questioning, how much time spends the particle inside the mountain?

An analogy to this would be an attempt to throw a ball so high that it escapes the gravitation forces of the earth and does not come back. To overcome gravitation means, in other words, to climb the gravitation mountain of the earth, the valley being the surface of the earth. To surmount the gravitation interaction an energy is needed, which is equal to at least a starting speed v_0 of 11.2 km/s (40 320 km/h), a speed called the *escape velocity*. A rocket which wants to reach Jupiter or Mars has to develop more than this speed. In other words, if this speed cannot be reached then the energy is too small and it is impossible to escape the gravitation of the earth. The ball or rocket fall back to earth or circle round it like a satellite. Instead of getting into the gravitation of Jupiter or Mars they fall back into the valley of the gravitation of the earth on its surface.

So, in the theory of quantum mechanics, there exists the tunnel effect, which indicates a distinct probability that there are particles that can get outside the energy valley, even though their energy is not sufficient to surmount the mountain.

It often takes many years until a particle, because of the negligible probability of tunneling leaves the valley of an atom nucleus. The carbon isotope ^{14}C for instance releases, on average, a particle from its nucleus after 5 600 years.

The radioactive decay of the carbon isotope ^{14}C helps us to determine very precisely the age of organic material. This was discovered by Willard F. Libby 1949 at the University of Chicago and in 1960 he received the Nobel prize in chemistry for his "*^{14}C method of determining age in archaeology, geology, geophysics and other sciences*".

This method functions like this: in the atmosphere carbon exists as three isotopes in different quantities, ^{12}C (98.89%), ^{13}C (1.11 %), and ^{14}C (0.000 000 000 01%). The index of the isotope reflects the total number of particles, protons and neutrons, because an atom can have a different number of neutrons in a nucleus. This differing number of neutrons with the same number of protons makes it possible for an atom to consist of nuclei of different size. These chemically identical atoms are called isotopes. In a carbon nucleus there are six protons. The atom is circled by six electrons. Isotope ^{12}C has six neutrons, isotope ^{13}C seven neutrons and isotope ^{14}C eight neutrons in addition to the six protons in the nucleus of the atom. (The number of protons determine the chemical properties of an atom).

Isotope ^{12}C and ^{13}C are stable, ^{14}C, however, is radioactive, which means unstable. The stability of an atom is determined by the composition of its nucleus. With isotope ^{14}C the eight neutrons cause instability compared to ^{12}C with only six neutrons.

All living things take in during their life ^{14}C isotopes through photosynthesis, breathing, or food. The ^{14}C concentration in living things is therefore physically balanced with the atmosphere. Whenever something dies no more carbon is consumed and the ^{14}C-concentration decreases through of radioactive decay at a constant rate (see Figure 5.9). By measuring the remaining activity of the ^{14}C carbon isotope the age can be determined. After 5 600 years only half of the original concentration of the ^{14}C isotope can be traced, after another 5 600 years only one quarter of the radiation is left. With the decay of the ^{14}C isotope in comparison with today's atmosphere the age of organic material can be determined up to 70 000 years. One of the triumphs of the ^{14}C method was the determination of the age of the Egyptian mummies.

Another tunneling process, which is vital for our life, has been happening for billions of years in the sun and in the

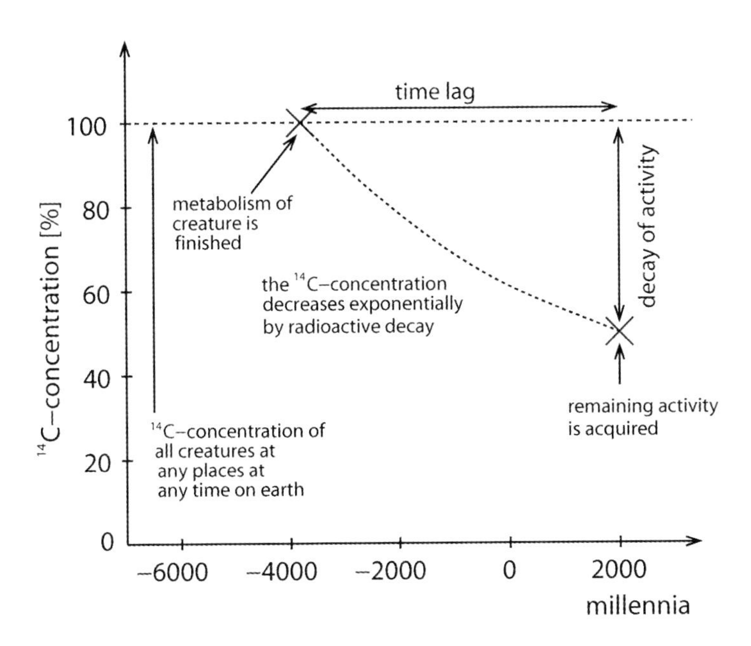

Fig. 5.9 Radioactive decay of ^{14}C isotopes allows one to determine the exact age of fossils.

stars – nuclear fusion. Nuclear fusion is an energy releasing process. It allows two atomic nuclei to melt into a new nucleus. This process creates the heat of the sun. This mechanism takes place at enormous temperatures, during which matter is in the fourth physical aggregate state, so-called plasma, which means atoms are in a state of such high kinetic energy they dismantle into their particles, electrons and nuclei. In the plasma electrons are separated from the nucleus, they are ionized. Electrons (negatively charged) and atom nuclei (positively charged ions) move at great speed independently of each other. The electrical attraction

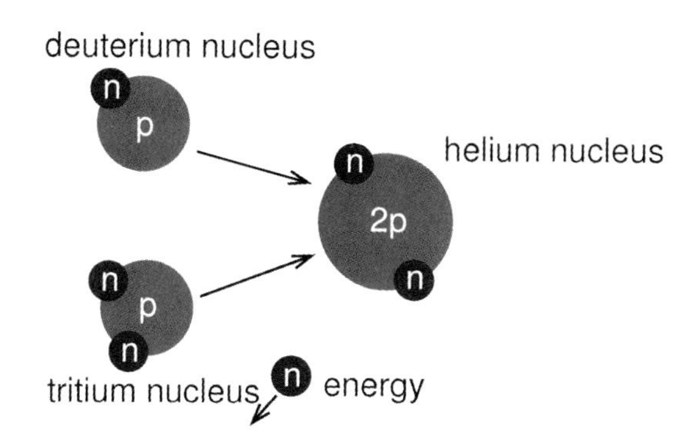

Fig. 5.10 Fusion of two hydrogen nuclei. When two protons p fuse one neutron n is released with enormous energy.

energy of the charged particles in plasma is smaller than their kinetic energy.

To fuse the two nuclei have to be close to each other so that the attracting nuclear forces are stronger than the Coulomb repulsion of the two positively charged nuclei, which acts as a high mountain. To overcome this repulsion the temperature would have to be 560 million degrees. Since the kinetic energy, which corresponds to the temperature, is only 15 million degrees inside the sun, it is not enough to overcome the Coulomb repulsion. The nuclei of heavy hydrogen, deuterium, can only fuse with the nuclei of supra-heavy hydrogen, tritium, to become a helium-4-nucleus, a so-called α-particle, through tunneling. The α-particles, also called nucleons, consist of two protons and two neutrons. With this fusion four times more energy is released than with the fission of a uranium nucleus. This energy explosion causes the already mentioned inconceivable force of the hydrogen bomb.

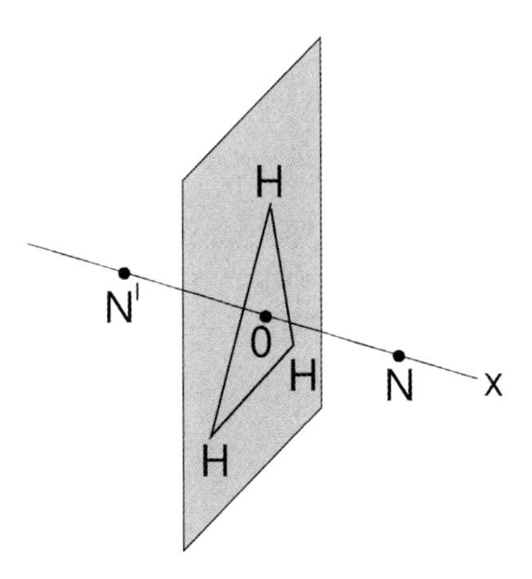

Fig. 5.11 Inversion motion of the nitrogen atom N to N' and vice versa in the same ammonia molecule.

Figure 5.10 shows the fusion of two hydrogen nuclei to form a helium nucleus (deuterium = one proton plus one neutron and tritium = one proton plus two neutrons). Deuterium ^{2}H and tritium ^{3}H are isotopes of hydrogen.

Penetration of potential barriers was already discussed with respect to molecules by Friedrich Hund around 1927, as mentioned in the Introduction. The ammonia molecule, NH_3, is a pyramid with the N atom at the vertex and the three hydrogen atoms H at the base, as shown in Figure 5.11. Obviously the N atom may be at one of the two symmetric equilibrium positions N and N′ on either side of the base of the pyramid [16]. Since both N and N′ must be equilibrium positions, the potential energy for the motion of the N atom along the axis of the pyramid must have two minima and have the symmetric shape indicated in

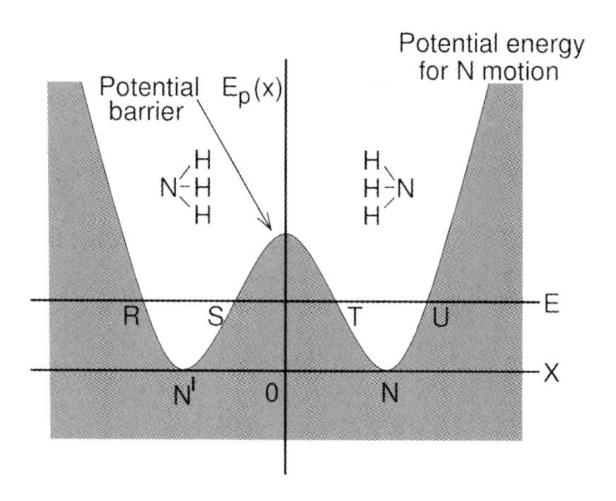

Fig. 5.12 Potential energy with the tunnel barrier for the inversion motion.

Figure 5.12 with a tunnel barrier between N and N′. If the N atom is initially at N, it may eventually leak through the potential barrier and appear at N'. If the energy of this motion is less than the height of the potential barrier, such as the energy level E in Figure 5.12, the motion of the N atom is composed of an oscillatory motion between R and S or between T and U, depending on which side of the plane it happens to be, plus a much slower oscillatory motion between the two classical regions passing through the potential barrier. Incidentally, the frequency of this second motion is 24 GHz. It is this second type of motion which we use to define a time standard with atomic clocks [16].

The tunnel effect has been applied technically since 1960 in semiconductor electronics in the Tunnel diode, called the Esaki diode after its discoverer. This diode is found today in many micro-electronic devices. Semiconductors are materials which are insulating at low temperatures but act as conductors at high temperatures or if they are doped with acceptors and donors.

Because of the close vicinity and periodic order of many identical atoms plus their electrons in the solid state, energetically close orbits of electrons come together and form so-called *energy bands*. Electronic orbits are then no longer precisely defined, they form a large band of energy orbits. Those orbits correspond to permitted bands of energy. Electrons, which move in these orbits can no longer be assigned to individual atoms but only to the complete semiconductor crystal. This phenomenon is also called an electron gas between the atoms of a crystal. The individual energy bands are, according to semiconductor material, separated by energy gaps of different size. The reason for the difference in conducting behavior between metals and semiconductors stems from the so-called *forbidden energy gap*, which separates the energy band completely occupied by electrons, *the valence band*, from the next higher energy band, *the conduction band*, (see Figure 5.13).

At low temperatures, i.e. small kinetic energy, electrons of semiconductor atoms can occupy the valence band completely. Thereby electrons cannot move as all spaces are occupied. With rising temperatures, however, more and more electrons get into the conduction band. On applying an electric field electrons can move in the conduction band, like in a gas, and also in the free spaces left in the valence band, called *holes*.

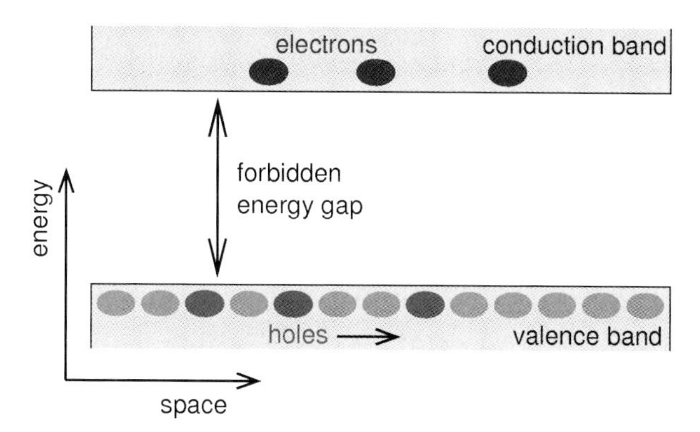

Fig. 5.13 Energy band model of a semiconductor. In the *forbidden energy gap* no electron can exist according to quantum mechanics. This gap corresponds to a mountain, which can be surmounted with sufficient thermal energy or by absorption of light particles. Otherwise it can be overcome only by tunneling. Only electrons with higher or lower energy than the energy gap can stay in a semiconductor.

The principle of a semiconductor can be explained with an analogy. Imagine a two-storey house with an entrance and an exit. The ground floor is filled with people carrying goods from the entrance to the exit. The ground floor is the valence band, the first floor is the conduction band and the people are the electrons. If there are only a few people on the ground floor then transport is no problem, the building acts like a good conductor. If the ground floor, however, is filled with people and the first floor is closed no transport is possible, the building is like a good insulator. If, however, the first floor is open for those who have got the energy to get to the exit in this way then transport is possible across both floors. To get through the first floor is no problem, and on the ground floor there is more space for transport. In the semiconductor the missing electrons correspond to the *holes* which also contribute to the conducting capability.

As mentioned before, some electrons acquire, from the thermal motion of the atoms, the necessary energy to get through the energy gap into the conduction band. In order to increase the conductivity semiconductors are doped with foreign atoms – impurities – which results in additional electrons (n-conduction, people on the first floor) or holes (p-conduction, less people on the ground floor). If an n-conducting semiconductor is connected to a p-conducting one, the interface of the two acts as a rectifying device, a rectifying diode. The current in such a structure depends on the applied voltage. The current–voltage characteristic is now not symmetrical. The current is higher e.g. for a negative voltage than for a positive voltage.[1]

Depending on the special doping of the two semiconductor device parts the energy-band structure can be designed as sketched in Figure 5.14(a). The energy is higher in the p-type material than in the n-type material. At the same energy the electrons in the conduction band are now oppo-

1) The current–voltage characteristic represents the dependence of the current on the applied voltage. If the current increases in proportion to the voltage the characteristic is called linear or *Ohmic*. If the current–voltage relationship is not linear the characteristic is called *nonlinear* or *non-Ohmic*. The magnitude of the current may also be dependent on the polarization of the applied voltage, such a characteristic is called asymmetrical. This behavior was discovered during the investigation of several minerals by Ferdinand Braun (1850–1918). Incidentally, Braun invented the famous electron beam tube used in classical TV-tubes and many other electronic devices.

site to the holes in the valence band. They are separated by a potential barrier, the forbidden energy gap. Even though the electron and the hole energies are too small to reach the positive conduction band or the negative valence band, respectively, the particles can tunnel through the forbidden energy gap.

This tunneling current causes the strongly nonlinear current–voltage characteristic useful for many technical applications (see Figure 5.14(b)).

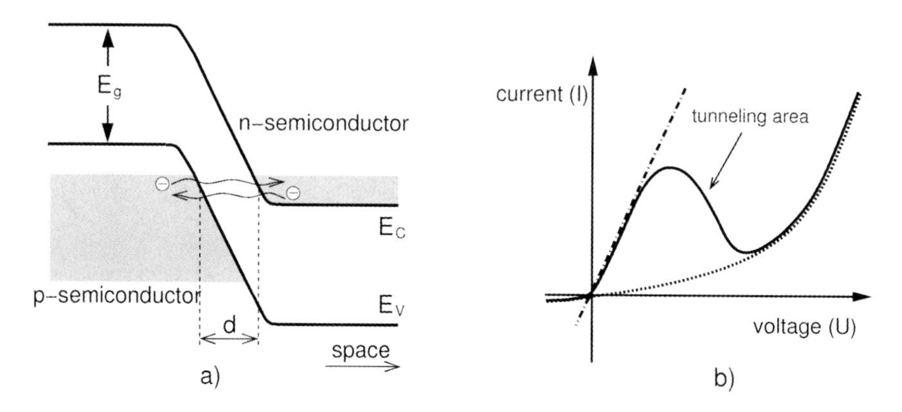

Fig. 5.14 (a) Energy band scheme at a semiconductor pn-junction. (b) Current–voltage characteristic of a tunnel diode. The dotted line represents the characteristic of the simple rectifying diode where the energy step at the junction is smaller and tunneling does not occur.

5.1.2 Tunneling Time

Above we have discussed the importance of tunneling for evolution and for applications. Tunneling is astonishing, mountains can be overcome without the necessary kinetic energy. The other exceptional property is the tunneling time. Quantum mechanics predicted an imaginary time, i.e. zero time. This property was important for basic research, especially for applied physics such as semiconductor tunneling devices or tunneling microscopy or signal processing in glass fiber optical systems.

The theoretical phase time approach introduced by Eugene Paul Wigner and David Eisenbud results in a zero time for the spreading of evanescent modes and the Schrödinger quantum mechanical equation yields the same result for the tunneling process. What is meant by an imaginary time and a zero time? Can the theoretical results be experimentally proved?

Quantum mechanical calculations by Thomas Hartman in 1962 pointed to such a purely imaginary, not measurable, not ascertainable, time in the tunnel barrier [17]. Until then, however, nobody had interpreted or taken seriously all the zero time calculations. Was the quantum theory correct with its prediction that there are spaces which could be crossed in an imaginary time, i.e. a time that cannot be measured by electrons, photons, atoms or even molecules?

Some years later there were published further quantum mechanical calculations indicating that evanescent modes are virtual photons which have a negative energy. Virtual photons violate the Einstein energy relation and can travel at any speed. The calculations are based on quantum electrodynamics (QED) [18, 19]. This famous theory of the interaction between light and matter was introduced in about 1948 by Richard Ph. Feynman, Julian Schwinger, and Sin-Itiro Tomonaga [20]. The virtual photons are necessary, for instance, to describe the interaction between two electrons. They exchange photons that never really appear in the initial or the final state of the experiment.

An analogous process takes place in photonic tunneling. The photonic barrier is bridged by virtual photons. The real incident and the real transmitted photons are tunneled by virtual photons.

The Feynman diagram of the tunneling process is sketched in Fig. 5.15. The virtual photon in this figure has not an arrow indicating his non locality and it is horizontally drawn demonstrating its space like nature. Virtual particles like tunneling photons are not observable and spreading in zero time as was shown for the case of evanescent modes by Stahlhofen and Nimtz recently [21].

Until then there was no empirical knowledge of the time spent in the tunnel. The experimental fixing of tunneling time of electrons in semi- and superconductor structures has failed to this day because of parasitic time losses. In semi- and superconductors there are chemical impurities and structural defects which cause time losses during the electron travel. The parasitic time losses fake a long tunneling time. The so-called semiconductor tunnel diodes have been used for a long time in microelectronics, but nobody

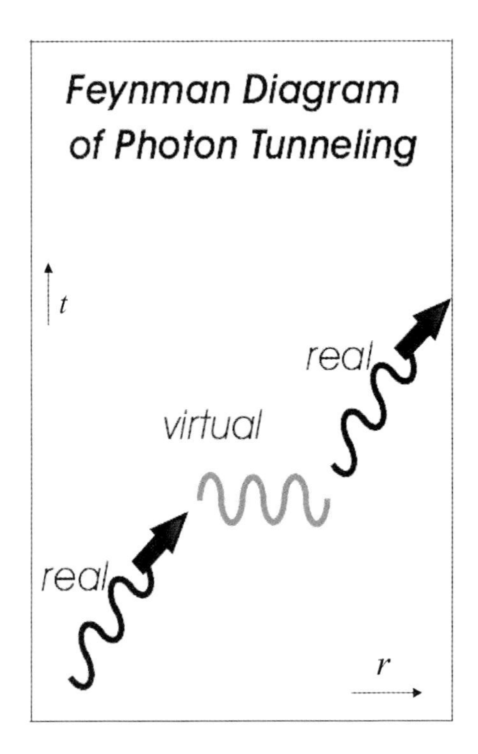

Fig. 5.15 Feynman diagram of the tunneling process. In the *barrier* no real photon can exist according to quantum mechanics. The link between the incoming and tunneled photon corresponds to a virtual, i.e. non observabel virtual, photon.

yet knows, how fast the electrons pass through the barrier, how fast the micro-electronic device can switch or oscillate ultimately. Recently, two Russian scientists Sekatskii and Letokhow succeeded in estimating, with the help of an electron microscope, the tunneling time of electrons. We shall come back to this first successful electron-tunneling-time experiment in due course.

Mathematics gives us a chance to get round experimental problems with electronic systems. It can be seen that mathematical descriptions of the tunneling of photons or α-particles are identical with the tunneling of photons, which means the tunneling of light particles and thus electromagnetic waves. In classical physics tunneling of light is described by so-called *evanescent modes*. Seen mathematically the results of measuring the tunneling time of photons should be identical with the tunneling time of electrons and other particles.

As mentioned before, photons can be described as light particles or waves or, more generally, as electromagnetic waves. Heat radiation, light, microwaves, X-rays or television and radio waves are all electromagnetic radiation, differing only in their wavelength and thus frequency. The wavelength λ of outgoing light waves or their photons is inversely proportional to the frequency f and thereby to their energy $E = h \cdot f = h \cdot c/\lambda$. Where h is the Planck constant and c the speed of light in a vacuum.

The energy quantum $E = h \cdot f$ is the smallest energy of a field with frequency f. This quantity is called a photon in the electromagnetic field. Therefore the energy of a field exists only as a multiple of this quantity.

The particles of electromagnetic radiation are stronger the higher the frequency f and the smaller the wavelength λ. For the speed of light in vacuum c: the simple relations hold as before, $f = c \cdot \lambda$ or $\lambda = c/f$. Figure 5.16 shows the spectrum of electromagnetic waves.

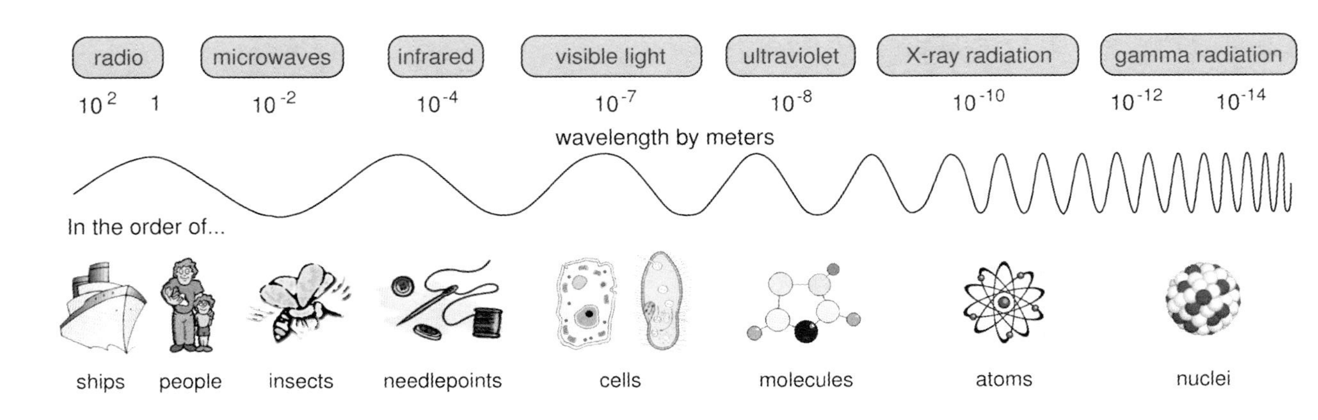

Fig. 5.16 The spectrum of electromagnetic waves. To compare the magnitudes of wavelength λ there are objects of every day life as well as those taken from biology and nuclear physics. Frequency f is calculated from $f = c\lambda$, based on the speed of light of $c = 300\,000\,000$ m/s. Thus a wavelength of 30 cm corresponds to a frequency of 10 MHz or to 10 000 000 oscillations/s.

Radio stations work with relatively low frequencies, in the kHz and MHz regime. The household voltage frequency is much lower, only 60 Hz. Radio waves are some 100 meters long. Microwaves, also called RADAR (**R**adio **D**etection **A**nd **R**anging, as used in air traffic control, a system used during the last world war and still today), have wavelengths in the centimeter region.[2] After the infrared radiation in the region of some micrometers, which we feel as warmth, there is, for us, the visible light with wavelengths of 0.4 μm and 0.7 μm. In this wavelength regime is the radiation maximum of the sun according to Wien's displacement law. The temperature of the surface of the sun is 6 000 °C. It seems that the human eye has adapted to this maximum radiation region of the sun. At still smaller wavelengths and higher frequencies one reaches ultraviolet radiation, X-rays, and, eventually, the highly energetic γ-radiation, which emerges when atoms decay.

For comparison we give the wavelengths of electrons. The wavelength of an electron depends on its kinetic energy and is only 0.001 μm with an acceleration voltage of one volt. High energy electrons, as used in modern accelerators with an acceleration voltage of around 100 GV (= Giga Volt; 1 Giga = 1 billion = 1 000 000 000) have wavelengths of far less than 10^{-17} m . With *light* of such a short wavelength microscopic structures in the region of 10^{-17} m could be observed and *optically* resolved. (An object can only *be seen* when it is larger than the wavelength of *light*. In the elec-

2) The household microwave has a frequency of 2.45 GHz. This corresponds to a wavelength of 12.2 cm. Satellite TV works with 10 GHz and a wavelength of 3 cm.

tron microscope an electron beam replaces the *light*). The positively charged proton, the element of the atomic nucleus, has a diameter of about 10^{-15} m, thus it is possible to examine the structure of a proton with such high energy electrons.

Tunneling experiments with photons in the microwave frequency regime have, compared to electron tunneling, several strong advantages. First: light particles possess no rest mass. They have very small energy quanta and can be produced and detected very easily. In addition they are not electrically charged so there is no interaction among themselves nor with other electrical charges of the environment. So, if tunneling photons with the frequency of microwaves are used instead of visible light, the photons have a comfortably large wavelength in the centimeter region, which makes measuring much easier. Measurements in centimeters can be carried out with higher precision than in the micrometer region. The high precision technique to conduct such microwave experiments to resolve time has been available for more than a decade.

Thus it became very attractive to use, instead of electron tunneling, analogous experiments to determine the tunneling time with microwaves. The results could then, through mathematical equivalence, be transferred to electrons and generally to all particles. As was mentioned in the preface, Italian physicists and engineers started such experiments in 1991.

5.2 Photonic Tunneling Structures

For electrons, mountains arise through nuclear forces, which are being tunneled while decaying. Repulsive electric Coulomb forces build mountains during nuclear fusion, less clearly, quantum mechanical effects of interference build up *forbidden energy gaps* in the semiconductor which have to be overcome. What do tunnel mountains for light particles look like? Figure 5.17 shows three well-known photonic tunneling barriers.

- Double Prisms:

  We start with the historically most important tunnel barrier in optics. Here, for tunneling, the principle of *frustrated* total reflection of wave beams is used. With this arrangement a small part of the reflected beam tunnels as evanescent modes through the air gap from the first into the second prism, although the incoming beam at the border of the first prism should be totally reflected. This is

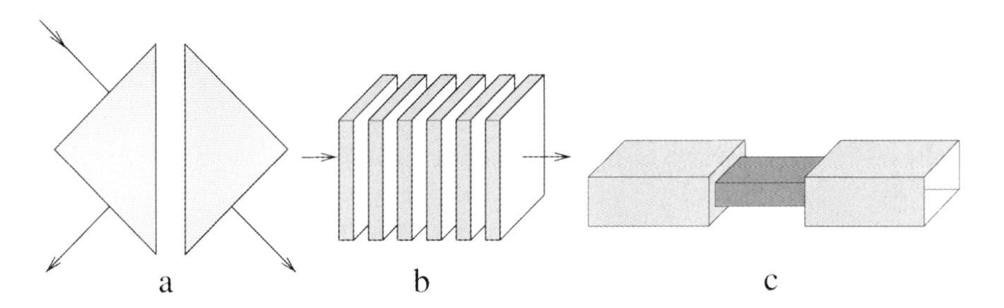

Fig. 5.17 Three different microwave photonic tunneling-barriers: (a) double prisms, (b) periodic lattice structure, (c) undersized hollow waveguide.

what *frustrated total reflection* means. If the second prism is removed the beam will be totally reflected at the border of the first prism. After that there will be a total reflection without frustration and tunneling.

- Photonic lattices ($\lambda/4$ lattices):

  Periodic dielectric quarter-wavelength structures are called $\lambda/4$ photonic lattices. This lattice structure consists of layers of different optical density. For instance the order air–glass–air–glass–air–glass–air can be chosen, whereby the thickness of the layers has to correspond to one quarter of the incoming wavelength. The passing waves then exactly disappear through overlapping. Waves which are reflected at single layers overlap constructively, they sum up to total reflection.

- Undersized wave guides:

  The wave guides here are hollow metallic pipe structures. These guides restrict the propagation of waves if they are too narrow. Wave guides are used frequently in microwave technology and are usually rectangular in shape. As soon as the cross-sections of the wave guides are smaller in both directions than half the wavelength of the transmitting wave, a wave can no longer exist, the wave guide becomes undersized. It can cover an undersized distance only through the tunneling process. In this case engineers also call the tunneling fields evanescent modes. Tunneling takes place at frequencies below the so-called cut-off frequency, below which regular wave propagation in the hollow waveguide is not possible.

All three structures have in common that they produce frequency windows in which normal wave propagation can no longer take place. Fields are spread as evanescent or tunneling modes only. These structures present to electromagnetic waves of certain frequencies or wavelengths, mountains, which can only be passed via the tunnel effect. In due course we shall explain the properties of tunneling barriers.

5.2.1
Double Prisms

Tunneling has been researched in classical optics for more than three centuries in the case of total reflection. This is a light phenomenon, like refraction and partial reflection, which can be experienced in every day life. Everybody knows different types of mirrors or has observed that a spoon in a cup or a stick in a lake seem to be bent, as sketched in Figure 5.18.

Fig. 5.18 Refraction of light: the spoon in the cup looks to be bent. *Photo: H. Kropf, HMI Berlin*

Total reflection happens when looking through the surface of water, as can be seen in Figure 5.19. Looking from under water against the light it is refracted. At an angle of $\alpha_{\text{total}} = 48.5^\circ$ the refracted light runs parallel to the surface of the water. At a larger angle light will be completely reflected into the water, there will be total reflection. The phenomenon is known as *fish eye effect*. The effect allows the human eye to enlarge its angle of view.

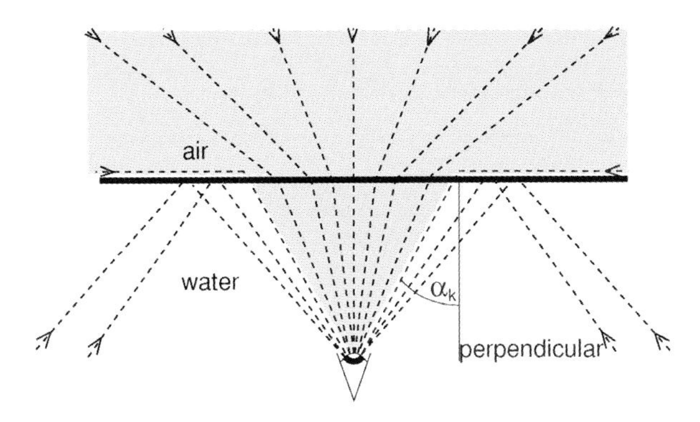

Fig. 5.19 Total reflection at an air–water interface. The illustration displays the light beams seen by an observer in the water. α_{total} is the angle of total reflection above which a beam cannot leave water or cannot enter from the air.

This effect is described by Snell's law of refraction and is illustrated in Figure 5.20. It means that light hitting two dielectrically different media, like water and air, will be partially reflected at the same angle ($\alpha = \alpha'$) and also refracted into the other medium. The electric properties of matter are described by the refractive index n. If the other medium is more dense, as when going from air into water light is re-

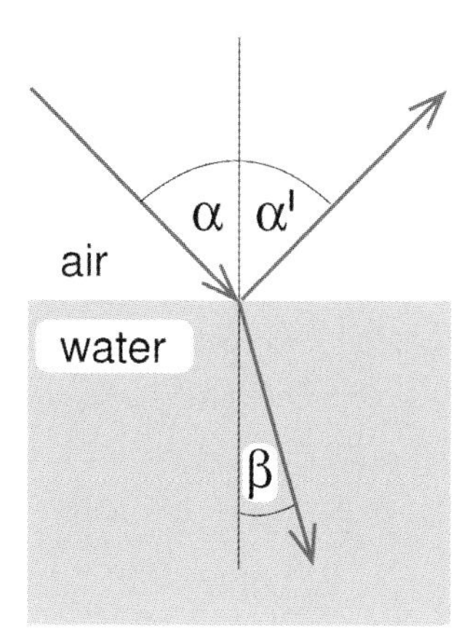

Fig. 5.20 An illustration of Snell's law of refraction.

fracted at a smaller angle, (see Figure 5.20). If light travels from the dense to the less dense medium, from water into air, as shown in Figure 5.19, it is refracted at larger angles. One realizes, that there is an *angle of total reflection* α_{total} above which light cannot, according to the laws of geometric optics, enter the second medium, it will be thrown back into the first medium. Snell's law of refraction goes like this: $n_1 \sin(\alpha) = n_2 \sin(\beta)$. β and α are the angles as shown in Figure 5.20, n_1 and n_2 the refractive indices of air and water. $\sin(\alpha)$ means one has to apply the sine function of α. For water and air the refractive indices are 1.33 and 1, respectively. For the angle α' of the reflected beam, $\alpha' = \alpha$.

Total reflection finds important technical application for instance in guiding light and infrared signals via glass fibers. Light fed into a fiber will be totally reflected from the surface of the glass fiber and thus, with small losses, transported through the cable (see Figure 5.21). Because of the total reflection the light cannot leave the glass fiber and therefore propagates only along the fiber.

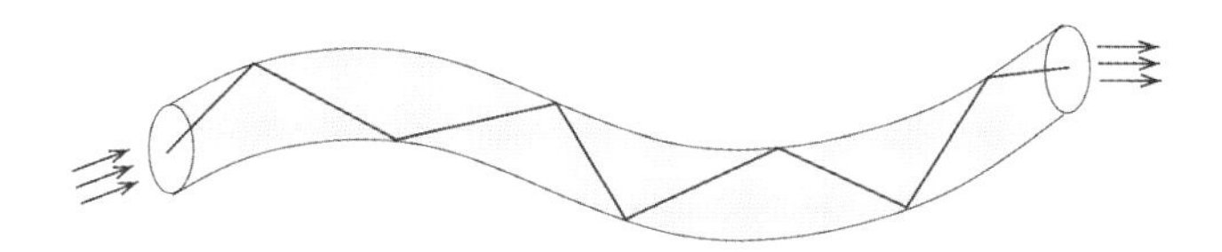

Fig. 5.21 Total reflection at the glass fiber surface guides the infrared light signal along the fiber, the latter being only some micrometers thick. As illustrated in this figure a glass fiber represents a light trough. The surrounding air acts as a repulsive potential barrier.

Until now we have described optical phenomena with geometric optics. To consider photonic lattices we have to move to wave optics. In reality the totally reflected beam penetrates a little into the less dense medium before it is reflected. It is interesting, and has been conjectured by Newton, that the totally reflected beam is shifted against the incoming ray. It is not reflected at the spot where it comes in but is moved by about a wavelength as was shown by Haibel and coworkers recently [22]. For light in the visible regime the shift is not possibly to see with the human eye because of its tiny wavelengths ($\approx 0,5\,\mu$m). With microwaves, however, the wavelength is some centimeters so that the shift of the wave beam by about a wavelength is clearly visible. This shift of radiation was first demonstrated by Fritz Goos and Hilda Hänchen in 1947. Since then it has been called the *Goos–Hänchen shift* [23]. Goos and Hänchen measured the shift of a reflected beam of glass with visible

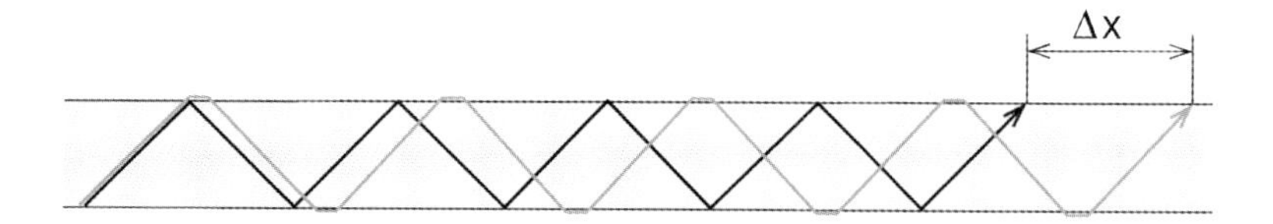

Fig. 5.22 Goos–Hänchen shift of a light beam along a glass slab. The sum of the individual shifts is given by Δx compared to a non-shifted beam. The shift was conjectured by Newton 300 years ago.

light (see Figure 5.22). To make the shift of the tiny wavelength visible they used a trick. They let the light be reflected frequently, so that the displacements added up to a large accumulation which was easy to observe.

Incidentally, it has been assumed until now that a shift between the incident and the reflected beams takes place only in the case of total reflection. However, quite recently it was observed that such a beam shift also happens in the case of partial reflection [24].

An experiment to show total reflection and the accompanying shift is the double prisms experiment, sketched in Figure 5.23 [22, 25].

Light- or microwaves entering vertically cross a double prism closed in the shape of a cube. When the double prism is opened the air gap between the two prisms behaves like a barrier against electromagnetic waves. The beam hits the border of the first prism at an angle larger than the total reflection in a medium with a lower refractive index. According to geometrical optics one expects a total reflection of the incoming beam (the dotted line in the figure). In reality the beam, as mentioned, penetrates the second medium,

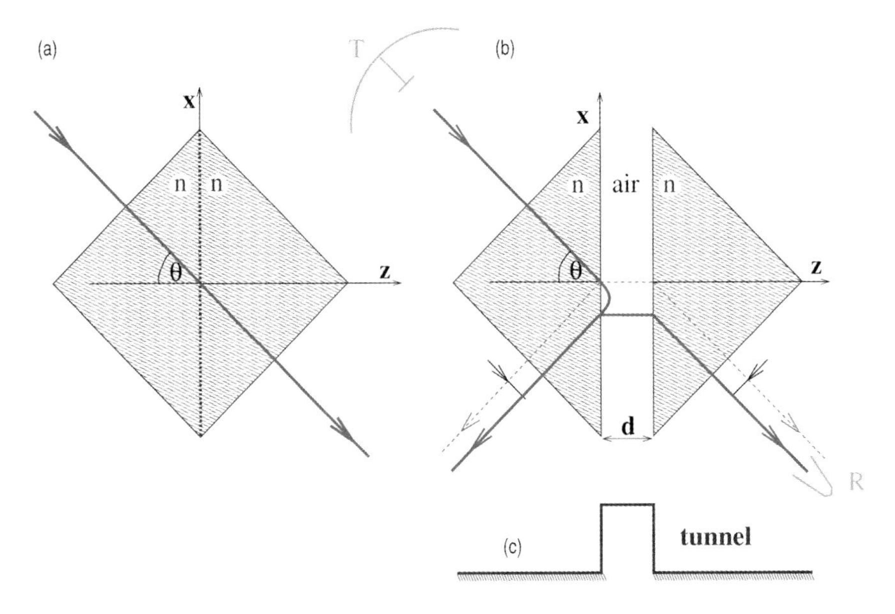

Fig. 5.23 Total reflection at double prisms. (a) The beam crosses the closed double prisms. (b) The air gap represents a tunnel barrier, as indicated in (c).

air, and moves a short distance along the surface before it is reflected (the continuous line in the figure). In this process a so-called surface wave emerges on the prism surface.

The wave number k, which describes the surface wave in the air gap mathematically separates into two parts, a real part $k_{\parallel}$ for the propagation along the prism surface and an imaginary part $k_{\perp}$, which describes the instantaneous vertical spread across the gap to the second prism and its fast fading. The intensity of the spreading of the latter field decreases exponentially with the distance to the surface of the prism (Figure 5.24). As the distance doubles the field strength fades to less than a tenth.

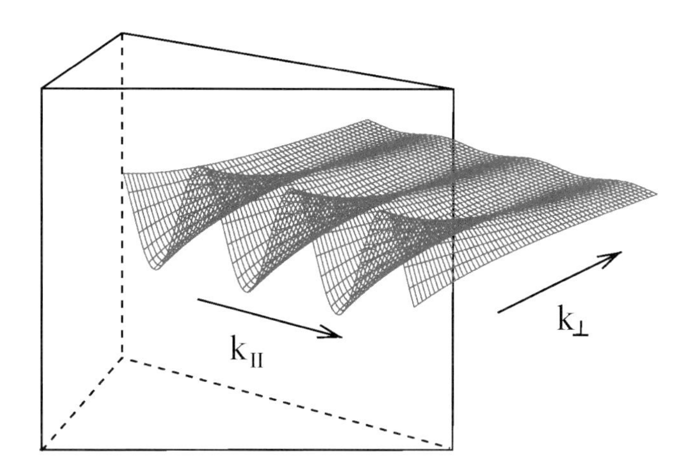

Fig. 5.24 Surface waves in the case of total reflection at double prisms. $k_{\parallel}$ shows the propagation direction along the surface and $k_{\perp}$ the direction perpendicular to the surface, the direction of tunneling.

This, with increasing distance, fast vanishing field is therefore called a vanishing or *evanescent* mode. It can be caught and measured by the second prism. Thereby the first prism is *frustrated*, the incoming wave packet is no longer completely reflected, because a small part gets through the tunnel, the air gap, into the second prism. The spreading of evanescent modes corresponds to the tunneling process of particles in quantum mechanics.

The evanescent modes, as they occur with total reflection are imaginary mathematical solutions of the law of refraction. Formerly they were rejected by scientists as non-physical because these solutions describe field modes with no real wavelength.

Fig. 5.25 Jagadish Chandra Bose at the Royal Institute in London, 1897 [26]. *Photo: IEEE, D. Emersion*

The fast exponential fading was proved by the Indian botanist Jagadish Chandra Bose (1858–1937) in his laboratory in 1897 with the help of high frequency waves.[3] Time measuring was not possible at that time, and the time behavior was not theoretically examined. Figures 5.25 and 5.26 show Bose in his laboratory and his instruments.

Measuring the time dependence of the tunneling process found no interest among physicists until fifteen years ago, although it posed exciting questions to examine the timelessness which had been predicted by quantum mechanics

3) J. C. Bose is not to be mixed up with Satyendra Nath Bose (1894–1974) who had predicted with Einstein the Bose–Einstein condensation of particles with a special spin. The predicted condensation of such particles was confirmed in 1995.

Fig. 5.26 Experimental set-up to measure the exponential decay of evanescent (tunneling) modes in the laboratory of J. C. Bose. The radio wave transmitter is on the left and the receiver on the right. In the center is a turntable with the double prisms [26]. *Photo: IEEE, D. Emersion*

in connection with the spreading of fields in tunneling barriers. Neither was the time behavior of the classical analogue, the evanescent modes in optics, studied. Only recently have experiments with microwaves tunneling the double prisms resulted in new experimental and theoretical discoveries in connection with the old and complex problem of total internal reflection and its tunneling properties [22]. The problems are of basic interest but are also important for many applications. Frustrated total reflection today is of great importance for instance in connection with determining the size of molecules in chemistry, measuring with tunneling microscopes, looking for biologically important molecules in medicine or, in optoelectronics, glass fiber signal guiding or signal transmission from one fiber guide to another by tunneling.

All these applications are based on the fact that the intensity of reflected radiation with frustrated total reflection reacts very sensitively to the optical properties of the material which is being examined. With this method of total reflection the refractive index of a material which has been deposited at the reflecting interface can be resolved with a relative precision up to 10^{-6}.

5.2.2 The Quarter Wavelength or $\lambda/4$ Lattice

Another way of constructing a tunneling barrier for microwaves and generally for photons or any wave packet is the $\lambda/4$ lattice structure. Such a structure is illustrated for electromagnetic waves in Figure 5.17(b). The effect of this structure is based, as mentioned before, on the extinguishing of overlapping waves, the destructive interference of waves.

This can be explained like this: whenever waves of the same frequency meet in the same place they interfere. So a wave emerges whose amplitude adds to the amplitudes of the other waves. This superposition of waves with identical frequencies is called *interference*.

The two extreme possibilities, *constructive and destructive interference* are shown in Figure 5.27.

If two waves meet in equal phases there will be a maximum of mutual strengthening, constructive interference, the fields add together. Equal phases means that the waves have the same displacement at the same time. If the overlapping is not equally phased they wipe each other out;

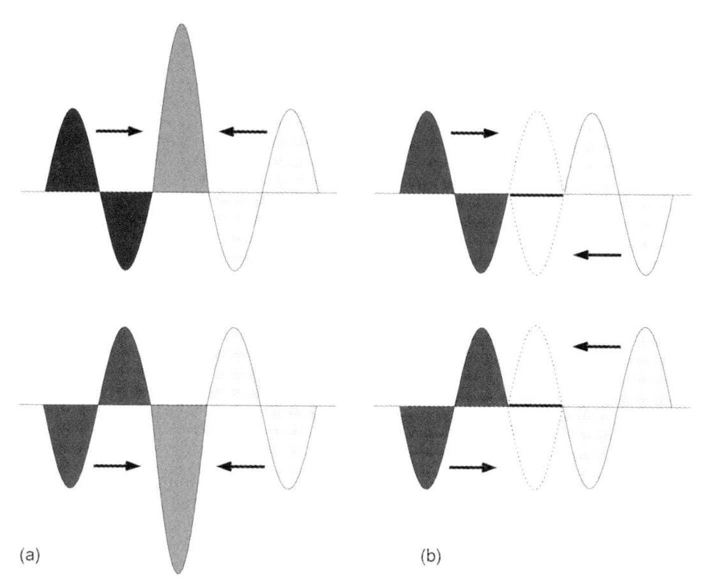

Fig. 5.27 Superposition of two waves of the same frequency. (a) Constructive interference. At the meeting point of the waves the amplitudes of both are orientated in the same direction and they add together. The amplitude is maximal. (b) Destructive interference. The amplitudes of the two waves are in opposite directions and they cancel each other out at the meeting point.

this is destructive interference. In most cases the resulting waves are between those two extremes. Waves can, according to their interference, strengthen, weaken or even annihilate each other. Thereby interferences do not annihilate or create energy, it is just a local reduction or increase in intensity. An example of interfering waves is shown in Figure 5.28, the interference of two waves of water.

The total annihilation by destructive interference corresponds to the tunneling wall designed with a $\lambda/4$ lattice. For such a lattice, distances are chosen so that photons of

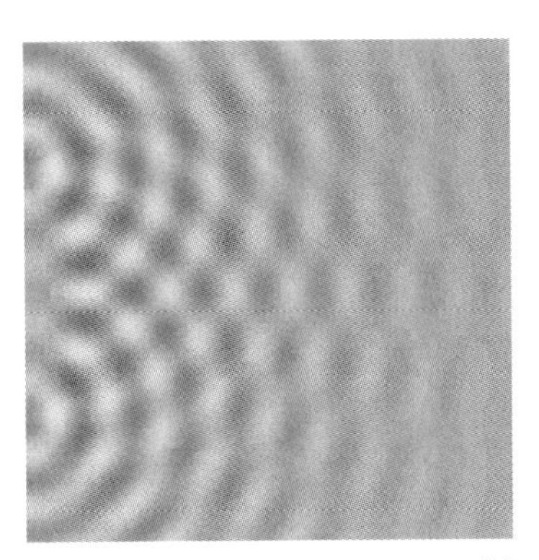

Fig. 5.28 Two interfering water waves. As they moved from their origin at some points they were both rippling up at the same time, at others they were canceling each other out.

wavelength λ can no longer propagate, apart from tunneling. The periodic distances have to be one quarter of a wavelength, a $\lambda/4$ of the photons.

Such photonic lattices can be built of glass whose planes are placed at a distance of a quarter of a wavelength. Each plane is transparent. Any incoming radiation is reflected by each plane – a little. The reflected beam suffers a change in propagation, so that with incoming radiation it causes a destructive interference. This repeats itself with each plane, so that at the end of the lattice there will be no radiation left, because the waves have extinguished each other during their transmission. The incoming beam has been reflected completely. If such a lattice were used for visible light then at the other end no light would arrive, it would stay dark: the lattice represents a perfect mirror.

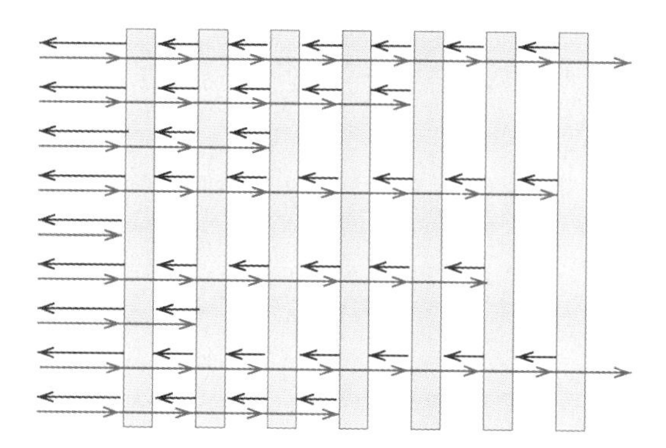

Fig. 5.29 Sketch of a $\lambda/4$ lattice with possible paths of incident and reflected beams of light. Most of the waves experience destructive interference and are reflected. Only a small amount are transmitted, by tunneling through the lattice. This principle is valid for electromagnetic waves as well as for electron waves and other particle waves. The lattice periodicity is $\lambda/2$, that means one grey plus one white $\lambda/4$ layer.

Actually, a small number of photons of *forbidden wavelength* get through the tunnel and reach the other end (see Figure 5.29). Their number diminishes with an increasing number of lattice panels. The analogy to quantum mechanical tunneling is the higher the mountain the fewer particles get through while with the lattice – the more layers the fewer waves arrive at the other end.

The above-mentioned semiconductor electronic device, the tunnel diodes, function, because of their periodic structure, as the already described optical $\lambda/4$ lattice with the forbidden energy bands. With the semiconductor, electron waves interfere with the atoms of the semiconductor because of their reflection by the atoms arranged at a distance of $2 \times \lambda/4$. The periodic structure forms the semiconductor–crystal lattice.

5.2.3 The Undersized Hollow Waveguide

In the high frequency region of electromagnetic waves for high power transport hollow wave conductors are mostly used. In such wave guides electromagnetic waves of high frequency propagate with little attenuation compared with other wave guides such as a coaxial line.

Coaxial transmission lines, when transmitting waves with frequencies of 1 GHz (1 GHz = 1 000 000 000 Hz) and higher suffer enormous attenuation, which increases considerably with higher frequencies. Furthermore, electric disruptive discharge may occur because of the small distances between internal and external cable conductors. At high power transmission this can lead to an electrical breakdown. Figure 5.30 shows a coaxial transmission line.

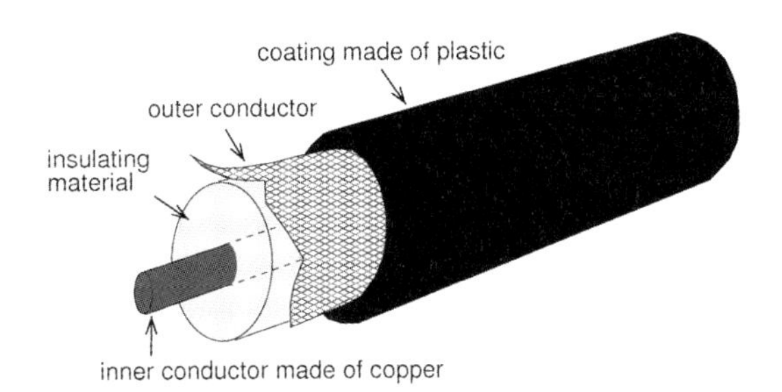

Fig. 5.30 A coaxial cable with an inner and an outer conductor separated by insulating material. The cable is coated on the outside.

The cause of the attenuation and breakdown is the insulating material which separates the central conductor, usually a wire or a thin hollow cylinder, from the outer cylindrical conductor or *shield*. All insulating materials, whether plastic or ceramic, absorb, with increasing frequencies, more and more electromagnetic power. To avoid these drawbacks hollow conductors are used in radar installations or microwave furnaces when strong high frequency signals are transmitted. Hollow conductors are metallic hollow bodies of round, elliptical or rectangular shape, which are filled with air or, for higher resistance against discharge, nitrogen, or are even evacuated. Figure 5.31 shows a rectangular wave guide which is generally preferred, the size of the guide being adapted to the special wavelength regime.

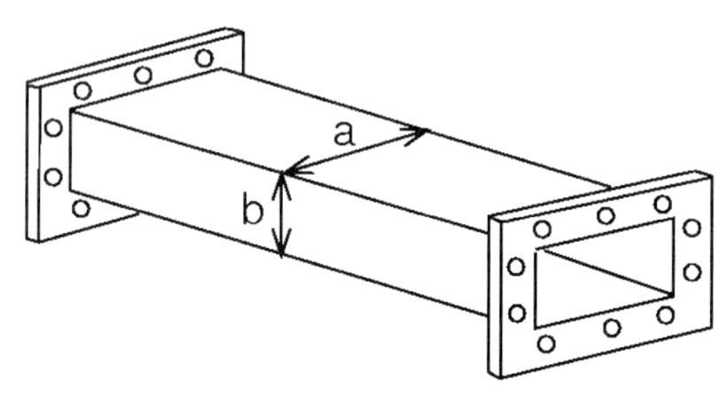

Fig. 5.31 An example of a hollow metal wave guide with a rectangular cross-section with flanges at the ends.

However, a wave can only propagate in a hollow conductor when its two transverse widths, especially that of the larger side a are at least half as long as the wavelength. If the wavelength is longer than half of both the widths of the hollow conductor no propagation can take place. The wave is then only able to *tunnel* through the too narrow conductor, as mentioned above.

To construct a tunneling barrier three hollow conductor pieces of different widths are put in line, as sketched in

Figure 5.32. Thus in the first and third sections a wave with a larger wavelength can propagate than in the middle of the set-up. The so-called cut-off wavelength, at which wave propagation is stopped, separates the regime of wave propagation from the tunneling regime of the hollow wave guide.

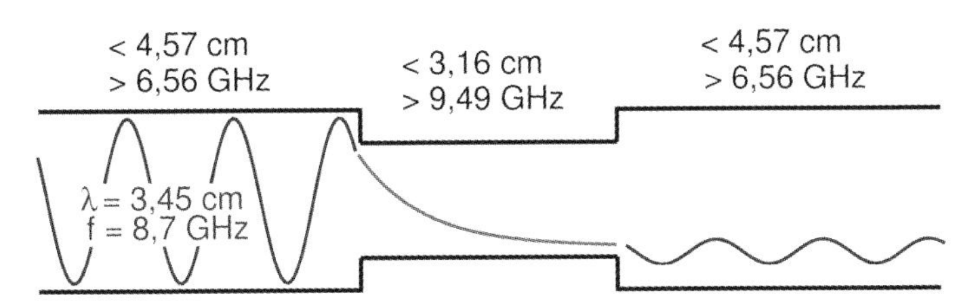

Fig. 5.32 Wave guide with an undersized center part, the tunnel barrier. The cut-off frequencies and wavelengths are given in the figure. The large and the small sections are called *X*– and *Ku*–band guides, respectively.

A signal whose wavelength lies between the cut-off wavelengths of the two different hollow conductors is introduced into the first section. Figure 5.32 shows a signal wave of wavelength 3.45 cm. This is longer than the cut-off wavelength of 3.16 cm in the middle section. In the first section the wave can propagate normally, in the middle section its wavelength is too large for propagation. The larger part of the wave is reflected but a smaller part can tunnel through the *forbidden* section as *evanescent* mode. At the end of the *forbidden section* there is the third section, which again allows normal propagation. The narrow section is a simple stretch of tunnel. It is interesting, that the tunneling field or particle in the tunnel cannot be measured. It is described by virtual photons with a negative energy. Measuring the tunneling photons (evanescent modes) is possible only outside the tunnel barrier in the area of normal propagation. This property of non-measurability can be seen by analogy with quantum mechanical tunneling. It shows that the apparently classical evanescent modes cannot be described by the classical Maxwell equations completely, even if these are a mathematical solution of the former. Instantaneously spreading non-local fields with a negative energy cannot be described by the Maxwell theory [27]. Non-local means the evanescent field is spread over the whole tunnel at the same time.

5.3 Tunneling Velocity

Several tunneling experiments have revealed considerably higher speeds than *c*, that of light in free space. The first experiments observing superluminal speeds were carried out with microwaves 1991 at the University of Cologne [4]. Later examinations with single photons (Berkeley University) and short light pulses (Technical University of Vienna) confirmed the superluminal tunneling of electromagnetic waves. Now more superluminal data in the microwave- and infrared frequency regimes have come from several laboratories. First applications of superluminal signal propagation in opto-electronics have been tested [28].

In 1994 Horst Aichmann and Günter Nimtz performed an experiment with microwaves in the laboratory of Hewlett Packard, in which Mozart's g-minor symphony number forty was tunneled on a microwave carrier through a hollow conductor at a speed 4.7 times the speed of light in vacuum.

Compared with a vacuum-distance the time lead in the short tunnel stretch of 114.2 mm was 300 ps (one picosecond (ps) = 10^{-12} s). The signal was detected at a speed higher than the speed of light, contrary to the naive interpretation of the special theory of relativity. This experiment was provoked by a claim by the physicists Th. Martin and R. Landauer. They had argued that superluminal speeds may happen, but they could not transmit information – information like a Mozart symphony [29].

Günter Nimtz performed at the Massachusetts Institute of Technology (MIT) the tunneled piece of music. Professor Francis Low, who organized the Seminar walked up and down in silence for some minutes, after that his only comment was: *This is not g-minor!*. The recorder had been running too fast and produced another key, which the professor with his perfect pitch had realized at once. Apart from that he and his guests were speechless for a moment, they had listened to superluminal music, i.e. to a *signal*, but they found it difficult to accept. Almost all of them thought, it would lead to violating Einstein causality. A very uncomfortable idea.

However, superluminal transmission does not violate the so-called primitive causality. Actually, the tunneling process only violates the special theory of relativity. Superluminal signals do not permit an exchange of cause and effect this means they do not violate the primitive causality. This problem will be dealt with in Section 5.5.

5.3.1
Measuring Tunneling Time with Double Prisms

Figure 5.33 shows the perspex double prisms, which were made from a diagonally cut cube of edge lengths $40\times40\times40\,\text{cm}^3$. The prisms are irradiated with microwave pulses (similar to what is shown in Figure 4.6) of frequency $f = 9.15\,\text{GHz}$ and pulse-half width of about 8 ns (1 ns = 1 nanosecond = 0.000 000 001 s). The refractive index of the perspex prisms is $n = 1.6$. For this value Snell's law gives an angle of total reflection of $\theta_t = 38.5^\circ$.

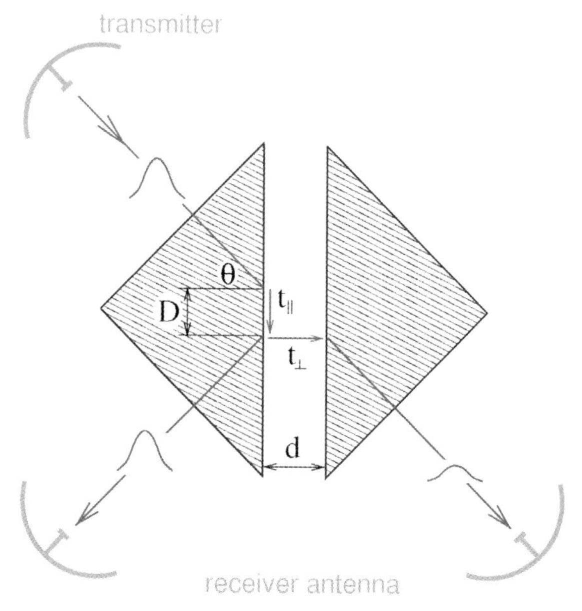

Fig. 5.33 Design of symmetrical beam paths for measuring the time to traverse the gap. $t_{\|}$ and $t_{\perp}$ represent the propagation times parallel and perpendicular to the prisms' surface.

The angle of the incoming beam and the first prism is $\theta_i = 45°$. This angle is considerably larger, by $6.5°$, than the critical angle θ_t of total reflection. The overwhelming part of the incoming beam will be reflected, a smaller part *frustrates* the reflection and tunnels, via the air gap, from the first to the second prism.

The time of the reflected as well as the transmitted signal can be measured. Comparing the times of the signal it is surprising that they are identical, independent of the width of the gap d. Since, with the symmetrical course of the beam, the reflection distance differs from the transmission by distance d, the latter must have crossed without time ($t_\perp = 0$!) as Figure 5.33 shows. This simple but logical experiment was carried out recently by the coauthor Haibel and coworkers [22].

The time $t_\parallel$ of propagation of the wave (see Section 5.2.1) along the surface of the first prism will be determined with a little trick. First, the propagation time of the signal reflected at the air gap is measured, then a mirror is positioned in the gap on the surface of the first prism. This reflects the incoming beam without any shift along this surface and without the frustration effect due to traversing the gap. Thus this measured time represents the propagation time of the reflected signal without a shift. From the measured time difference of the two procedures the time, $t_\parallel$, that the wave spent along the surface can be determined. In the experiment in question this time was 100 ps ($= 10^{-10}$ s). The whole tunneling time, $\tau = t_\parallel + t_\perp$, is 100 ps, whereby obviously $t_\perp = 0$. No time is needed to traverse the (tunneling) gap.

The evanescent modes traveled with an infinite speed inside the barrier.

If the air gap between the two prisms is fixed at a width of 5 cm a signal with the speed of light would pass the distance $D + d$ in a time $t_{\text{light}} = t_{\parallel} + t_{\perp} = 100\,\text{ps} + 165\,\text{ps} = 265\,\text{ps}$. The tunneled pulse reaches the end of the tunnel 165 ps faster, because in the tunnel $t_{\perp} = 0$. This is, over the distance $D + d$, a speed 2.65 times faster than the speed of light c. This experiment at the double prisms gap with a symmetrical beam shows simply but clearly that superluminal speeds are possible in tunneling.

5.3.2 Measuring Tunneling Time with the Quarter Wavelength Lattice

Figure 5.34 shows how tunneling time is measured with a $\lambda/4$ lattice. The photonic lattice, a periodic dielectric hetero structure consists, in this experiment, of perspex sheets separated by an air gap. For the microwave frequency 9.15 GHz the optical thickness of the sheets (0.52 cm $\times$ 1.6 refractive index) as well as the distance between them (0.83 cm) corresponds to exactly one quarter of the wavelength ($\lambda = 3.3$ cm) of the signal which is to be tunneled.

On the basis of this periodic structure the microwave signals interfere destructively, in other words the wave wipes itself out moving forward, it is thrown back. The lattice acts like a mirror, as explained above. This annihilation, i.e. destructive interference of a forward moving wave leads to a

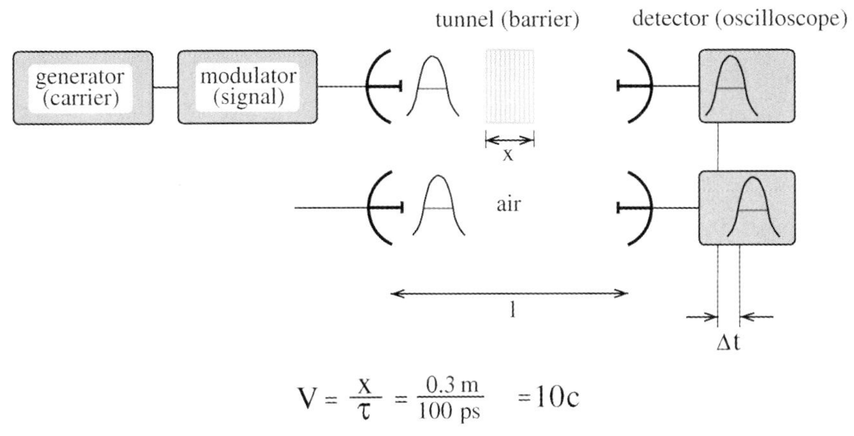

Fig. 5.34 Experimental set-up to measure the tunneling time. The propagation time through a barrier is compared with that over the same distance in air. x, τ, v, and c represent the barrier length, tunneling time, tunneling velocity, and velocity of light in vacuum.

strong fading or, mathematically, to an exponential decay with increasing lattice length. This is formally described by an imaginary wavenumber. Such a photonic lattice forms a tunneling barrier (as discussed in Section 5.2.2).

With the experimental set-up shown in Figure 5.34 the time for a signal to traverse a distance in air and the time to traverse the same distance with the tunneling barrier in place are measured. In this experiment the tunnel length is 30 cm. A microwave generator produces the carrier frequency, a modulator forms the individual pulses with a half width of about 8 ns (see Figure 5.35). The signals will be transmitted and detected by parabolic antennas, *bowls* as we know them in connection with television. A parabolic

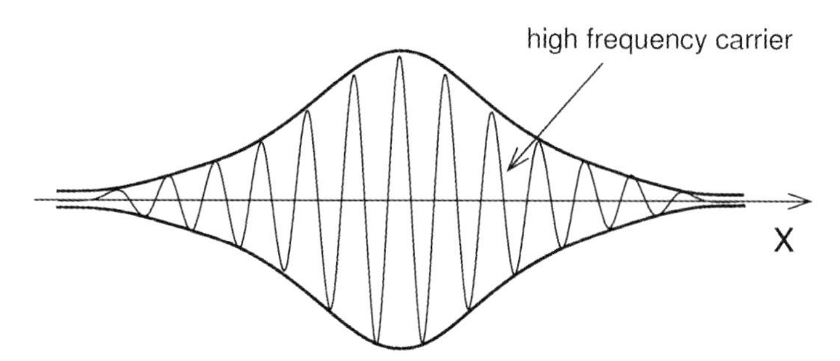

Fig. 5.35 Sketch of the envelope of a wave packet and the carrier frequency are indicated. The wave packet is representative of a microwave pulse as well as of electrons or other particles.

antenna radiates a parallel directed beam of waves. In this way the wave beam keeps its shape over long distances.

The propagation time of a digital signal, its pulse, is detected with an oscilloscope and compared with the time of the same pulse which has traversed the same distance through air.

The tunneling signal takes 100 ps to traverse a tunnel distance of 30 cm, the signal traveling with the speed of light takes 1000 ps, i.e. ten times longer. The intensity of the tunneled pulse is reduced to 1% of its incoming intensity, the signal traversing the air distance suffers no attenuation. Since the information of a signal is contained in its modulation, here in its half width modulation, and this half width is independent of magnitude it remains unchanged (see Figure 5.36). The intensity of the detected signal is of no importance for a signal as long as it is above noise level. Whether a melody is played forte or pianissimo does not have an effect on the information. So the tunneling signal will be detected

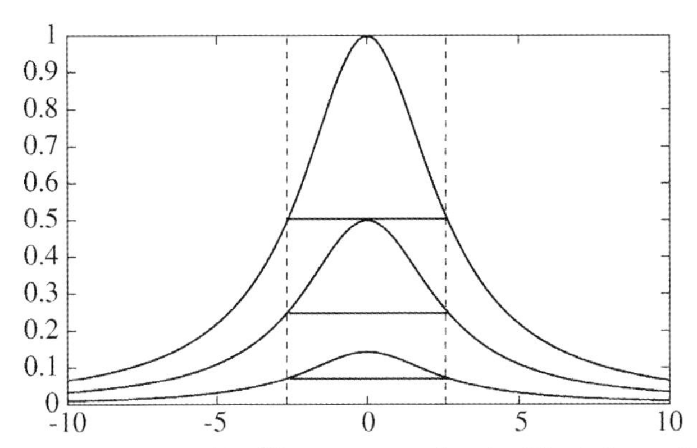

Fig. 5.36 The same pulse with different power levels. The information is the half width, which obviously does not depend on power.

900 ps sooner, which means it travels ten times faster than light propagated in vacuum.

Figure 4.6 shows digital signals of our contemporary communication systems. They are carried by infrared waves. Some years ago, in a laboratory of the Corning optoelectronic company in Milan, Stefano Longhi and his colleagues tunneled infrared digital signals of this kind with twice the speed of light along a fiber barrier [30].

5.3.3 Determining Tunneling Time with an Undersized Hollow Waveguide

To determine tunneling time in undersized wave guides the propagation times of microwave pulses are compared (see Figure 5.32) with the time the pulses lost traversing the same air distance. The carrier frequency of the pulses lies between the cut-off wavelengths of the hollow waveguide sections. To cover a tunnel distance of 10 cm it takes a microwave pulse with a frequency of 8.7 GHz 130 ps. At the speed of light it takes 333 ps to cover the same distance. The tunneling signal is again faster than light [4].

5.3.4 Tunneling: Zero Time in the Tunnel Barrier

Analysis of the data from our experiments shows that photons spend some time at the tunnel entrance and then cross the barrier in zero time. That no time is lost in the tunnel barrier had been predicted already by Thomas Hartman in 1962 on the basis of quantum mechanics [17] and later for electron tunneling by Francis Low and Peter Mende [31]. In 1991 these quantum mechanical conjectures were confirmed in analogous experiments by Achim Enders and Günter Nimtz [4]. This result can be deduced both theoretically and experimentally: the tunneling time does not depend on the barrier length. If the length of the barrier is doubled the time remains the same. This property is called the Hartman effect.

Because of the zero time inside a barrier the speed of a particle is infinite, it is omnipresent in this space. Only at the tunnel entrance is some time lost for photons or any other particle. This constant time, independent of tunnel length, leads to the fact that with a longer tunnel the tunneling velocity increases proportionally, because speed is length divided by time.

Therefore signals can be transmitted with superluminal velocity, even if they lose energy through reflection at the tunnel entrance [7]. This does not apply only to photons and electrons but also to atoms and molecules.

Fascinating evidence that macroscopic structures possess quantum mechanical qualities was produced by the team of Anton Zeilinger in 1999 at the University of Vienna [32]. It was proven that even molecules consisting of 60 carbon atoms, like *buckminsterfullerenes*, possess wave qualities (see Figure 5.37). These spherical-shaped molecules were named after the American architect Richard Buckminster Fuller (1895–1983) and are also called *football molecules*.

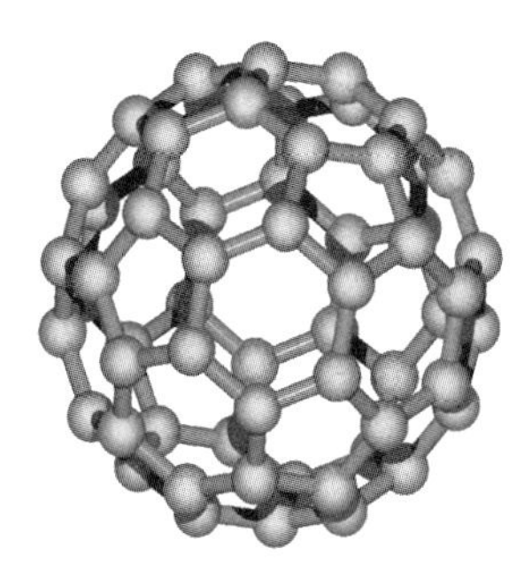

Fig. 5.37 Buckminsterfullerene molecule. The rather large football-like molecule is built of 60 carbon atoms. *By courtesy of W. Harneit, FU + HMI Berlin*

That these molecules behave like wave packets could be shown through interference measurements with a small lattice. When many of those molecules are shot against a grating an interference structure emerges, which allows the conclusion that the individual molecules have gone through at least two slits of the grating at the same time (see Figure 5.38).

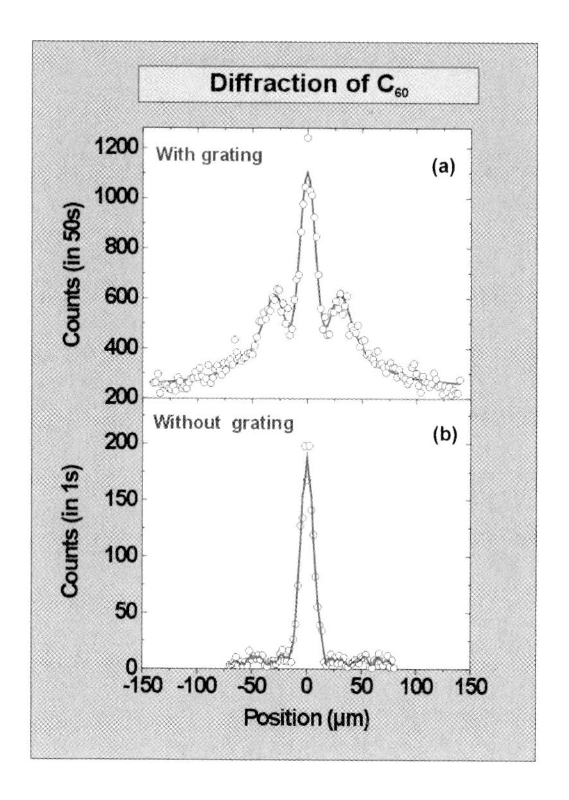

Fig. 5.38 (a) C_{60} molecules are refracted at a microscopic lattice. Obviously the molecule waves interfere in a pattern that we are used to from light. (b) The molecules intensity distribution measured at the same place without the lattice. The interference pattern has disappeared. *Homepage AG Prof. Zeiliger, Universität Wien*

These wave properties make it possible, theoretically and experimentally, to tunnel such big molecules. This means, that quantum mechanics moves into the regime of macroscopic objects. Originally physicists had assumed that quantum mechanics applied only to the microscopic regime of an atom and below.

5.4 Tunneling as a Near-Field Phenomenon

There is no doubt that tunneling takes place in *zero time*, the evanescent field is instantaneously spread over the barrier. However, we have seen that tunneling is an evanescent process, that means the probability of traversing a barrier decreases rapidly with the tunneling barrier length. The tunneling process is observable only over a limited barrier length, physicists call such a phenomenon a *near-field* or a *near-zone effect*. This property was exploited recently at the XXVIIIth Spanish Relativity Meeting by the coauthor Günter Nimtz [15]. A signal becomes a signal only by its defined effect. It has to be properly detected, which means the energy of the signal has to be above the receiver's sensitivity. A receiver specification is limited by the signal-to-noise power ratio, this ratio has to exceed 1 in order to measure a signal. As an effect of the temperature the detecting elements of a receiver are excited according to the corresponding motion of the electrons, which is called the noise power. Temperature is a measure of the kinetic energy of a particle. The voltage equivalent to the temperature was measured first by Johnson in 1928 and theoretically elaborated by Nyquist at the same time. The electrical noise power is given simply by the product of the detector temperature and the frequency band of the signal in question.

Thus, in order to detect a signal its power must exceed the power of the thermal noise. For the tunneling process, detection is limited by this fundamental physical property to a near-zone. In the case of microwave signals the near-zone can extend up to kilometers, and up to meters in the case of infrared signals. In the case of electrons or α-particles, however, the near-field zone has microscopic extension.

An important technical near-field effect represents the pattern of a dipole antenna radiation. Close to the dipole antenna magnetic and electric field components of the radio wave are traveling one after the other, in the far-zone after some wavelengths, however, both components are traveling parallel to each other but pointing in perpendicular directions.

5.5 Causality

Superluminal signal velocities lead, in the opinion of many popular books and text books, to very strange consequences of the special theory of relativity. For example: Calculations based on signal speeds higher than the velocity of light result under certain conditions in a chronological exchange of cause and effect [6]. For an observer, under certain conditions, the effect can appear *before* the cause. Figure 5.39 shows such a theoretical example [33].

Maria (A) transmits the lotto numbers to her friend Susanne (B) from position $x = 0$ at time $t = 0$. Susanne is located at time t' in a space ship at position x' and travels with a speed v_r of $0.75 \cdot c$ away from Maria's position x. For the transmission Maria uses a tunnel, which can transmit at four times the speed of light. Suppose the transmission distance L, the tunnel, between the two is 2 000 000 km, which is 50 times the circumference of the earth. Susanne now uses

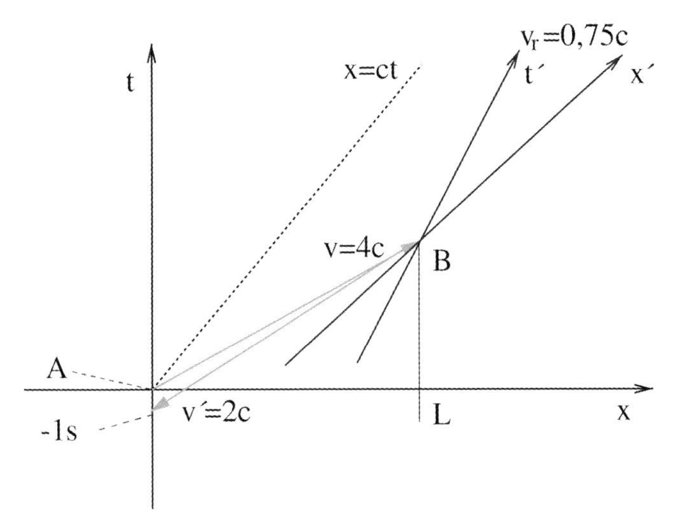

Fig. 5.39 Time coordinates of two inertial observers A $(0,0)$ and B with $O(x,t)$ and $O'(x',t')$ moving with a relative velocity of $0.75{\cdot}c$. The distance L between A and B is 2 000 000 km. A makes use of a signal velocity $v_S = 4{\cdot}c$ and B makes use of $v'_S = 2{\cdot}c$ (in the sketch $v \equiv v_S$). The numbers in the example are chosen arbitrarily. The signal returns -1 s in the past at A.

a tunnel of the same length to return the lotto numbers to Maria. With the formalism of the theory of special relativity it could be calculated (see Figure 5.39) that Maria would get the numbers back in the past (negative time values, before being sent off) so that the figures could be presented in good time (1 s) at the counter. Susanne gets her money.

With a naive interpretation of Einstein's causality, which excludes signal speeds higher than the speed of light, this change of cause and effect would be impossible. The experiments presented above have shown that superluminal velocities are indeed possible and can transmit signals and thus information.

The special theory of relativity does not describe tunneling, which does not take place either in free space or by normal wave propagation.

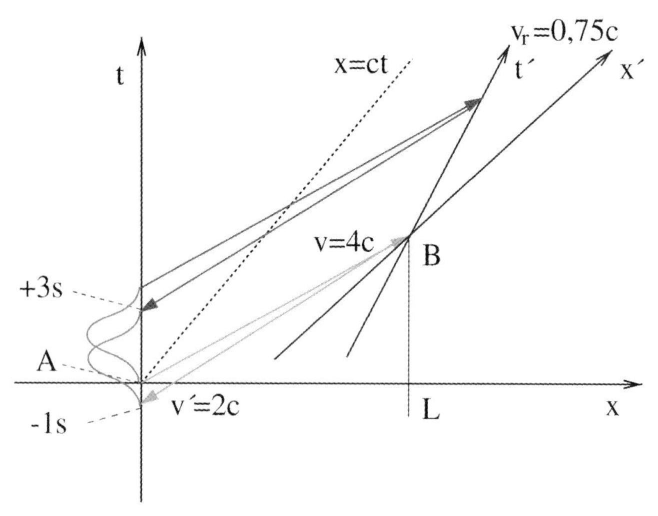

Fig. 5.40 In contrast to Figure 5.39 the pulse-like signal now has a finite duration of 4 s. This data is used to give a clear demonstration of the effect. In all superluminal experiments, the signal length is long compared with the measured negative time shift. In this sketch the signal envelope ends in the future by 3 s (in the sketch $v \equiv v_S$).

However, despite the superluminal signal speed in the tunnel and other superluminal processes the principle of causality, *cause before effect*, remains intact (see Figure 5.40) [15]. The explanation is that in Figure 5.39 the lotto figures are *non-physically* presented in *zero time*. However, a signal has a duration of time, some seconds are necessary, we assumed four seconds, to transmit the lotto figures or rather pronounce them. So we have to wait for the end of the signal in order to catch all of the numbers before we can hurry to the lotto counter. This finite time duration makes it impossible to manipulate the past. The beginning of the signal reaches system A earlier in the past but the end of the signal always arrives in the future of system A with positive time values, no matter what the signal speed was [15].

If one tries to outwit the principle of causality by letting the expansion of a signal get extremely narrow, so that it lies totally in the past, one would need an extremely broad frequency band in order to produce such a signal.[4]

These extremely short signals and therefore signals of wide frequency bands cannot be used for a non-causal application because they would suffer *dispersion* in the tunnel. Dispersion in this case means, that there are different frequency components of a signal in the tunnel, which are differently attenuated and differently delayed at the barrier entrance. High frequency components of a tunneled signal are attenuated in a different way to those of low frequency, which leads to a reshaping of wide band signals. If the envelope of a signal is reshaped significantly it can no longer be identified.

There is another reason as shown in Section 5.2.3, why a signal with a too wide frequency band can no longer tunnel. This is shown in Figure 5.32 with the undersized hollow wave guide. Through this hollow conductor a signal can only tunnel if its frequency band lies between the two cut-

4) This procedure is applied in modern communication systems in order to transmit more signals at the same time across the same transmission line. However, the frequency shift is only some orders of magnitude and not unlimited, an unlimited frequency shift would unlimit the signal energy.

off frequencies of the hollow conductors. If the frequency band gets too wide, the signal over- or under-steps these barriers. Overstepping the upper limit of the frequency then the wave fits into the narrow piece of the hollow conductor and propagates from there with the speed of light or at a lower speed. The narrowed undersized hollow conductor would then no longer be a tunnel. Falling short of the lower frequency components a wave can no longer extend in the wider wave guide.

Finally, and decisively, a wide frequency band corresponds to a very high energy, because energy increases with the frequency of a signal. This is shown in Figure 5.16: The minimal quantum energy of a microwave is smaller by billions than that of an X-ray. A signal containing such high frequencies has correspondingly high energies.[5]

5.6 Non-Locality: Reflection at Tunneling Barriers

Non-locality means that a wave package, a single physical particle or several entangled particles are spread across a given space at the same time (instantaneously). Theology assumes that God is omnipresent. From the physicist's point of view this God would be an example of a non-local phenomenon. Ordinary objects like human beings will be met at a given time at a given place, they are localized. A photon or electron can be met at the same time in New York

5) Despite this fact, in the literature signals are often non-physically described as of unlimited frequency band width and therefore of unlimited energy, (see for instance Ref. [6]).

or Paris. Only when measured will particles be localized, they are *arrested*, the wavefunction of the particle collapses.

A non-local behavior can be observed when a microwave signal is reflected at a tunnel entrance. The experimental apparatus to prove the non-local behavior of the tunnel process is shown in Figure 5.41.

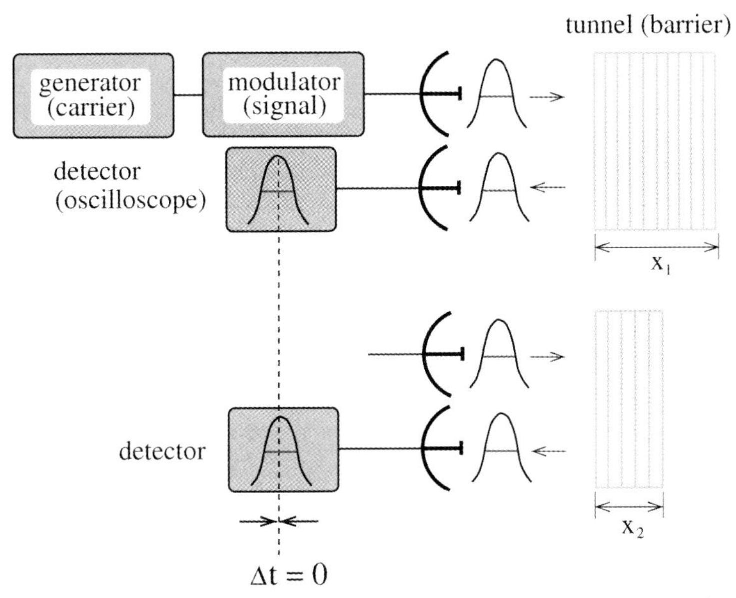

Fig. 5.41 Set-up to demonstrate non-locality by measuring the time dependence of partial reflection at a photonic barrier with a digital pulse. The parabolic antenna at the top of the illustration transmits digital pulses towards the barrier, the second one below receives the reflected signal. The time delay is measured with the oscilloscope.

A pulse-shaped signal was transmitted from a parabolic antenna through a variable-length tunneling barrier. The barrier was a $\lambda/4$ lattice. The reflected signal was received with a second antenna. The propagation time and intensity were compared with a signal reflected by a metal mirror. In place of a tunnel the mirror was placed first at the position of the tunnel entrance and then at the position of the tunnel end. The, from the classical point of view unexpected, result is shown in Figure 5.42.

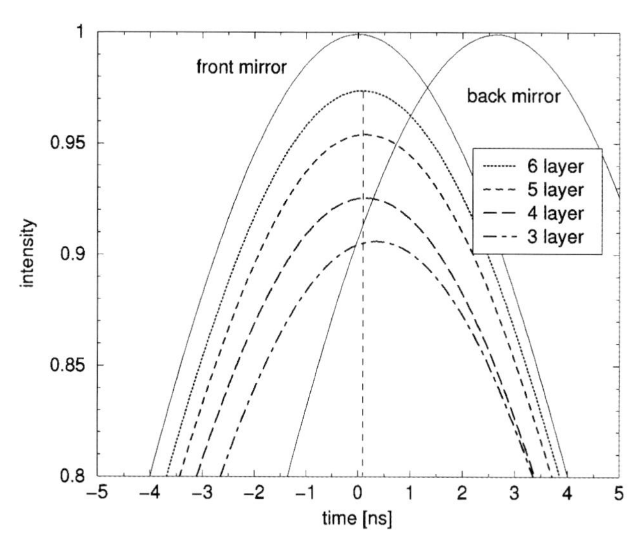

Fig. 5.42 Measured partially reflected microwave pulses vs time. The parameter is the barrier composition, as illustrated in Figure 5.41. The signal reflections from metal mirrors replacing the tunnel and placed either at the position of the tunnel entrance or exit are displayed. In this experiment the wavelength was 3.28 cm and the barrier length was 40 cm. The number of lattice layers was reduced from 6 to 3 inside the resonant lattice structure illustrated in Figure 5.41.

The signal reflected by the first mirror propagates in the air with the speed of light. This propagation time serves for calibration, its time duration was fixed at $t = 0$. If the mirror is placed at the position of the tunnel end the signal propagation time through air back and forth (twice the barrier length 2×41 cm) was 2733 ps.

The signal reflected at the tunnel barrier needs, however, only 100 ps, or the equivalent of one oscillation time of the wave more than that signal reflected by the mirror at the entrance from there to the detector, no matter how long the tunneling barrier. This time difference corresponds to the tunneling time of the barrier in transmission. It is interesting, that the tunneling time and reflection time are equal. Both come into being only at the entrance of the tunnel, before the signal spreads instantaneously inside the barrier and is transmitted and reflected. This means the time duration is independent of the tunnel length: all signals reflected at barriers of different lengths hit the detector in an equal time. This property was first discovered by Hartman when calculating the tunneling time and is called the Hartman effect, as mentioned before. However, with a shorter tunnel length the reflected part of the wave is reduced and more of the signal is transmitted through the tunnel. Figure 5.41 shows that with shorter length the reflected intensity is reduced from 97% to about 95%. With this simple microwave experiment the length of the tunnel barrier can be determined from the amplitude within a time of only 100 ps (100 picosecond = 0.000 000 000 1 s).[6]

6) Two linked (also called *entangled*) particles are supposed to have a positive and a negative spin. Thereby the total value of the spin is zero. If a positive spin of a particle is measured instantaneously the other particle is negative, no matter where it is positioned. A similar link can be found when tunneling reflected or transmitted radiation. The intensities of reflection R and transmission T add up to a normalized total intensity of 1: $(R + T = 1)$. If, at the entrance of the tunnel barrier, the reflected intensity R is known then the intensity T at the exit is known and vice versa. In quantum mechanics the combination $R + T = 1$ corresponds to the conservation of the number of particles, whereas in classical physics this is based on energy conservation. According to the theory of relativity we speak of a *space like*, i.e. a superluminal link of the quantities R and T with the total value 1, whereby R and T are positioned separately in the tunnel as shown in Figure 5.39.

So the wave package spreads in the tunnel in zero time and is everywhere from the entrance to the exit. This non-local phenomenon makes one feel eery.

The corresponding propagation time through air of 2×41 cm to the mirror at the end of the tunnel and back is more than 27 times longer, i.e. about 2 733 ps longer than the delay at the entrance.

Neglecting the near-field properties of the tunneling process the non-local reflection leads to a fascinating Gedanken (thought) experiment: if with the help of frustrated total reflection at the prism we look into the universe, for instance at the distance from the prism to the moon or to a star far away (all these objects are characterized by a change in the refractive index), we could determine within only 100 ps the distance to these objects through the strength of the reflection, the magnitude of our signal. Supposing, however, that we could measure the reflection factor to hundreds of decimal places.

5.7
Tunneling Particles are Not Observable

There is another strange property. Tunneling particles or any evanescent wave packet are not observable, that means detectable, inside a barrier. This is because inside a barrier their energy is formally negative, the particles and waves are not real but of *virtual nature*, as mentioned above. They do not interact with a detector or a receiver. They are observable only outside the barrier when converted back to a real particle or wave. Inside a barrier the particle could be evidenced only if there was added so much energy from outside that the tunneling process was annihilated and the total energy of the particles and waves, including the added energy, exceeded the barrier's height. A proof of the non-observability of tunneling particles can also be shown by the uncertainty relation [21].

5.8 Universal Relation between Tunneling Time and Signal or Particle Frequency

The analysis of tunnel experiments has led to an interesting result: The tunneling time τ approximately corresponds to the oscillation time $T = 1/f$ of the tunneling wave, which equals the reciprocal frequency (see Fig. 5.43). If there is a digital pulse, i.e. an amplitude modulated signal (AM), then the tunneling time τ is given by the reciprocal of the mod-

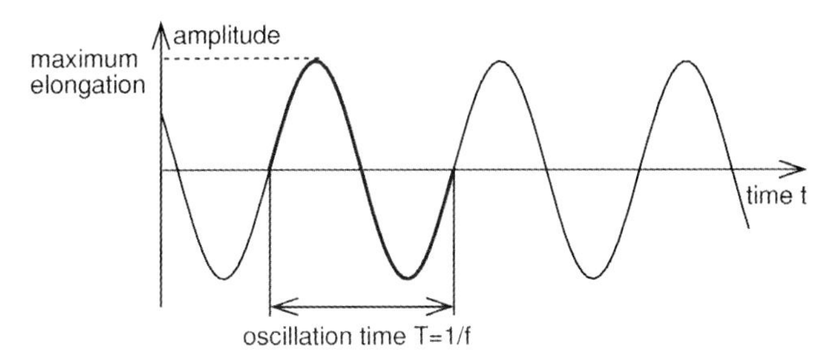

Fig. 5.43 Oscillation time T of a wave. It equals the reciprocal frequency f. An oscillator with the frequency 1 Hz has an oscillation time of 1 second.

ulated carrier frequency. In the case of a photon or another particle, on the basis of wave–particle dualism, the tunneling time is the reciprocal frequency $1/f$ of the photon or, in the case of any wave package with energy E (Figure 5.35) the reciprocal particle energy divided by the Planck constant h $(T = 1/f = h/E)$. This property is also called, after the French physicist and Nobel prize winner Louis Victor Prince de Broglie (1892–1987), the *De Broglie frequency*. In 1923 he proposed the hypothesis of wave–particle dualism, that is that light has simultaneously a wave-like and a matter nature.

Table 5.1 shows tunneling time data from experiments with photons measured by different teams of physicists with different tunnel barriers and with different frequencies [34]. As mentioned before, the tunneling time is given by the delay time at the tunnel entrance and no time is spent in the tunnel. Tunnel time depends only on the frequency or energy of the signal and not on the kind of barrier.

On account of the above mentioned mathematical analogy, the universal connection between tunnel time and reciprocal frequency of photons should also be valid for electrons, for α-particles, and for all other physical particles. This universal relationship thus allows, for the first time, the determination of the minimal response time of a tunnel diode solely by the tunnel process, without parasitical time-consuming interactions in a semiconductor.

The validity of the predicition of a universal tunnel time also for electrons was confirmed recently with a smart experiment. Two Russian scientists, Sekatskii and Letokhow, at the Institute of Spectroscopy in Moscow measured the tunneling time of electrons with a field emission microscope. The results are between 6 fs and 8 fs, according to the applied field strength (1 fs = 1 femtosecond = 0.000 000 000 000 001 s). The universal empirical connection with the tunnel time was estimated at $\tau > 2.43$ fs. A study on electron tunneling in semiconductor hetero structures by Pereyra yielded data also in agreement with the universal tunneling time as displayed in Table 5.1.

Electrons, like photons, possess particle as well as wave qualities. An electron looks, in the wave picture of quantum mechanics, as shown in Figure 5.35. From the energy of an electron its wave package frequency and hence its oscillation time can be deduced. Oscillation times and tunneling times of electrons measured in the field emission microscope correspond within the measuring accuracy. This up to date experiment confirms the assumption that electrons and even phonons (sound particles), like photons, spend no time in the tunnel [31, 35]. Like photons, electrons stay at the entrance during the tunneling process.

Recently the universal tunneling time was also observed in acoustic experiments, by Robertson et al. and by Yang et al. [36, 37] for example. The quantized particles of sound are called phonons. The acoustic experiments were carried out in a 1-dimensional sound tunneling tube analogous to undersized electromagnetic wave guides and in the forbidden frequency band of a 3-dimensional acoustic lattice structure. In both acoustic tunneling experiments the measured tunneling time equals approximately the inverse oscillation time of the sound pulses as shown in Table 5.1.

Tab. 5.1 Data of the photonic tunneling time measured with three different barriers and at very different frequencies. For comparison electronic and acoustic data are shown, too.

Photonic barrier	Reference	Tunneling time τ	Reciprocal frequency $T = 1/f$
frustrated total reflection at double prism	Haibel/Nimtz	117 ps	120 ps
	Carey et al.	≈ 1 ps	3 ps
	Balcou/Dutriaux	40 fs	11.3 fs
	Mugnai et al.	134 ps	100 ps
photonic lattice	Steinberg et al.	2.13 fs	2.34 fs
	Spielmann et al.	2.7 fs	2.7 fs
	Nimtz et al.	81 ps	115 ps
undersized waveguide	Enders/Nimtz	130 ps	115 ps
electron tunneling (field emission microscopy, semiconductor AlGaAs-GaAs)	Sekatskii/Letokhov	6–8 fs	>2.43 fs
	Pereyra	100 fs	37.5 fs
acoustic (phonon) tunneling	Yang et al.	0.6–1 μs	1 μs
	Robertson et al.	0.9 ms	1.12 ms

5.9 Teleportation

The term teleportation appears frequently in Science Fiction. It describes the possibility of converting a man to a pulse of electromagnetic radiation and send him to some distant planet in another galaxy. There, there are smart creatures, who reconstruct him with the instructions coded in the electromagnetic pulse using matter from their planet. Such a process is allowed by the laws of physics (of course, the technical implementation is impossible for the time being), however, the journey would be made at the speed of light and not superluminal, therefore it would take a long time to arrive at the planet of another galaxy. In quantum physics teleportation of states like electromagnetic wave polarization and particle spin orientation is possible, but again, the process proceeds not faster than the speed of light.

There is, however, a popular process of teleportation that has been used for a long time: the telephone. In this example sound is converted to electromagnetic waves, which travel about a million times faster than sound. Finally, the electromagnetic waves are converted back into sound at the receiver.

5.10
Wormholes and Warp Drives

Science Fiction deals again and again with the phenomenon of *time travel*. As long as we can travel only at speeds considerably less than that of light there is no chance of leaving our galaxy. The diameter of our galaxy is about 120 000 light years. Even if a space ship could travel at the speed of light it would take 4 000 generations before it would get from one end of the galaxy to the other. Could a tunneling process in the universe lead to a superluminal speed of matter? In the place of photons, electrons and α-particles space ships could tunnel through the universe to reach far away destinations or travel back into the past. In the areas of electromagnetic and nuclear forces tunneling is established, does that also apply to the area of gravitation which rules the universe?

Since Einstein's general theory of relativity (1915) the universe is defined by the four-dimensional *space–time structure*. In this structure beside the three space-directions there is time as an equal quantity. Time corresponds to a three-dimensional extension. The space–time structure is therefore four-dimensional.

As early as 1935, Albert Einstein and Nathan Rosen had proved that the general theory of relativity allows so-called *bridges* or *abbreviations*, which are called wormholes.

Normal matter possesses, without exception, positive energy and causes positive space–time curvatures, which we perceive as gravitation. To construct sufficiently large and stable wormholes it needs a negative curvature of space–time, a repulsing gravitation, a *negative* energy or mass, also known as *exotic matter*. Negative energy distorts space–time and opens astonishing possibilities. How then can negative energy be produced? We have learned that evanescent modes and, in general, tunneling particles are represented by virtual particles. They have a negative energy and they violate the relativistic energy relation.

On the other hand, a space area can, according to popular ideas in an absolute vacuum at best possess zero energy. According to the principle of uncertainty of quantum physics there are space areas which contain *less than none*, which is negative energy. In an absolute vacuum the average energy is zero, but so-called virtual oppositely charged particle pairs constantly arise and destroy each other. The density of energy fluctuates.

To create such negative energy density the so-called Casimir effect can be used. In 1948, the Dutch physicist Hendrik B. G. Casimir calculated that two uncharged, parallel-positioned metal plates could influence vacuum fluctuations so that the plates attract each other (see Figure 5.44). Whereas in the vacuum outside the parallel plates all sorts of fluctuations or wavelengths are possible, between the two metal plates only even numbered multiples of certain wavelengths can exist. Charge and energy fluctuations between the two plates will thus be reduced. From the outside more waves press against the plates than there is resistance from the inside. Negative energy develops, negative pressure pulls the plates together. The more the plates are pulled together the more the negative energy and negative pres-

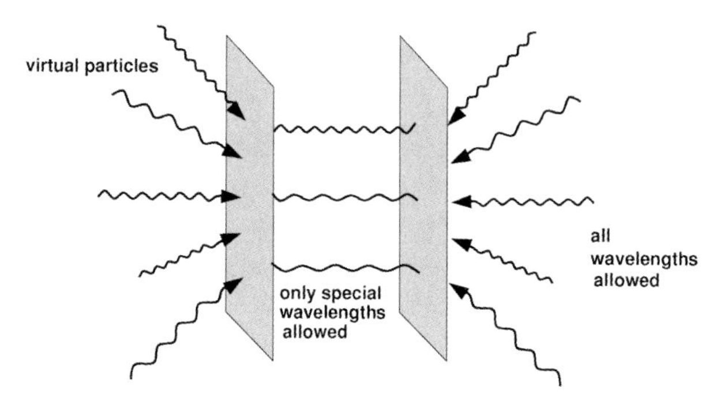

Fig. 5.44 Illustration of the Casimir effect. There are more wave packets outside the parallel metal plates and thus the sheets are pressed together.

sure increase and at the same time the strength of attraction increases. However, the effect is tiny. A plate distance of only 10^{-6} mm causes a negative pressure of just 10^{-4} of the air pressure assuming plates of size 1 square meter.

Quantum optics experts could prove that vacuum fluctuations can be suppressed by destructive quantum interference and thereby alternate areas of negative and positive energy are created. On average the total energy stays positive. Whenever in one place negative energy is created, then in another place positive energy must emerge. A pulse of negative energy is always followed by a pulse of positive energy, which overcompensates the former. This effect is called the *"quantum interest"*. Negative energy has to be paid back by positive energy and with interest (see Figure 5.45).

If negative energy can be produced, a very negative bent region in space–time could be constructed and thus a sort of superluminal drive, a wormhole that can be crossed, as

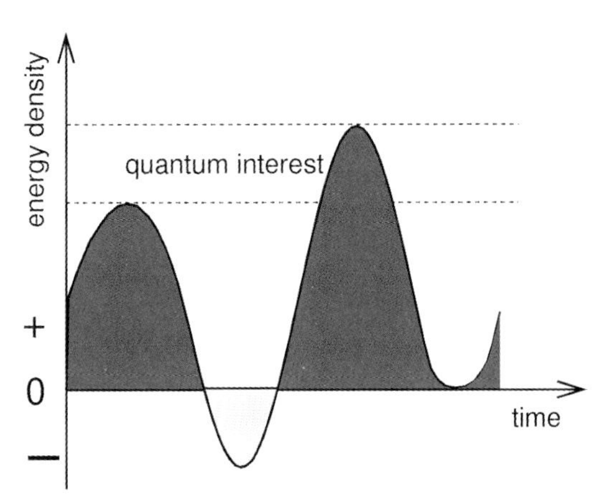

Fig. 5.45 Illustration of the quantum interest. The negative energy density produced at a place corresponds to the loan of energy to be paid back at another place with additional interest.

shown in Figure 5.46. The negative energy causes repulsive gravitation and prevents the collapse of the wormhole. The tunnel-shaped connection between two space–time areas makes it possible for particles or space ships to shorten time and distance between two points in the universe drastically and reach the destination faster than along the normal route. One could enter the wormhole on earth and land after a short time on the whirlpool-galaxy.

Macroscopic wormholes can be designed, but the negative energy would be limited to an extremely thin band round the hole. With a wormhole radius of one meter the thickness of a surrounding negative energy field would be 10^{-21} m. The necessary negative energy for the construction corresponds to the total energy which is produced by 10 billion stars within one year. If a wormhole were constructed

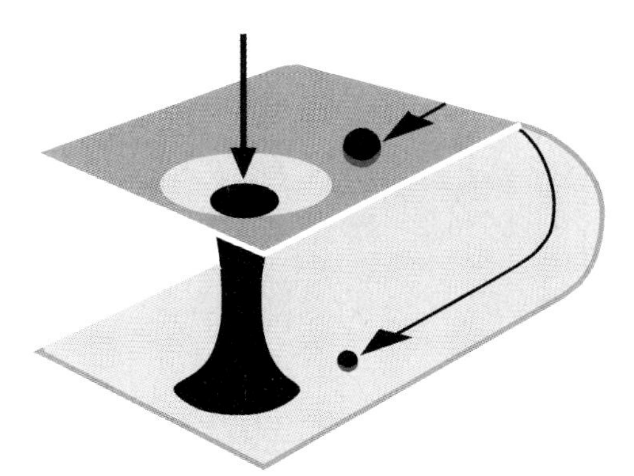

Fig. 5.46 A wormhole represents a tunnel-like shortening between two distant places in space. The tape represents the real space with the positive energy of our universe.

through which a space ship could travel, the total energy of the universe would not suffice.

Another possibility of superluminal space travel was suggested in 1994 by Miguel Alcubierre Moya of the University of Wales in Cardiff: the space–time bubble or warp drive. Figure 5.47 shows a space ship which travels at a speed faster than light in a space–time bubble. To construct such a space–time bubble negative energy is again needed.

The space ship rests relative to its surrounding in the space–time bubble, for observers outside it moves with the bubble at any high speed. The superluminal movement of the bubble can be achieved by a condensing of the space–time structure at the bubble front and an expansion at the back of the bubble. This shortens the distance to the space ship whereas the distance to the place of origin expands. This condensation and extension and the connected accel-

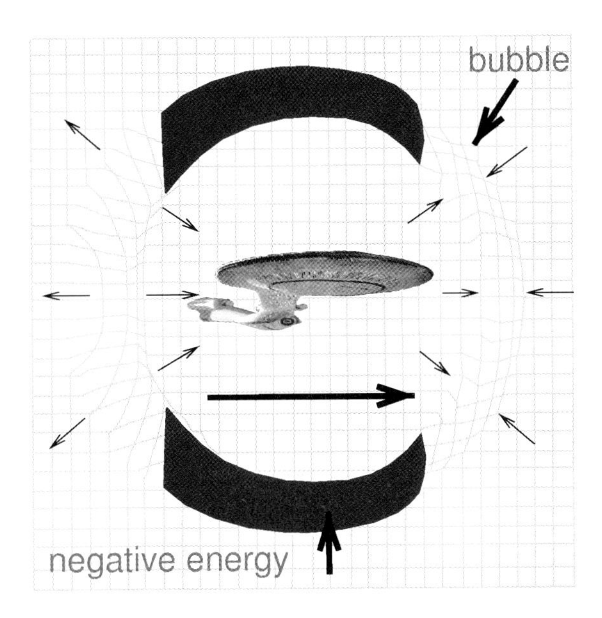

Fig. 5.47 The science fiction-like warp drive for space ships. Well known from the space ship Enterprise.

erated movement is possible according to the general theory of relativity. At the moment this explains the superluminal expansion of the universe. The space–time bubble corresponds therefore, like the wormhole, to the abbreviation in the four-dimensional space–time world.

Apart from the necessary energy to build the bubble there is, however, another problem of the warp drive. Between the outer fringe and the inside of the bubble there exists no causal connection. A space ship could not steer the surrounding bubble, it would have to be programmed from the outside before the start of the trip to get on the right course.

Of course nobody believed such a technical achievement as landing on the moon was possible only ten years before the first landing on the moon in 1969. The landing on the moon was possible according to the then prevailing state of physical knowledge. Even though the dream of flying to the moon came true, the superluminal trip through the universe remains an exciting intellectual game, which fails because of the gigantic amount of energy that would be needed [38].

6
Summary

Time is measured experience, recognizable for example in the swing of a pendulum, the orbit of Jupiter's moons or the length of a journey. Feeling and thinking also need time. This can be measured by an electrical encephalogram and by computer tomography. Physiologists observed that man needs some amount of time for feeling and thinking. We have called it man's time quantum which is a time unit of about a tenth of a second. Thus, on average, a human being has a supply of 40 billion human time quanta, which means he has 40 billion experiences or perceptions.

The propagation of effects and thus of signals takes place at a finite speed in free space, which is the speed of light, 300 000 km/s. Experimental and theoretical research in recent years have shown, however, that there are special spaces, tunnels in mountains and wormholes, which are crossed instantaneously. A particle, a space ship or simply information does not stay in the tunnel although it travels through it. Consequently, in tunnels or wormholes particles travel infinitely fast. Even so, the tunneling particle takes some time to get to the other end of the tunnel. This is be-

Zero Time Space: How Quantum Tunneling Broke the Light Speed Barrier. G. Nimtz and A. Haibel

ISBN: 978-3-527-40735-4

cause particles or wave packages interact for a certain time at the entrance of the tunnel before they either turn back or traverse it. The total process of tunneling leads to velocities many times faster than the speed of light. This again allows a shortening of the time that passes between cause and effect compared with the speed of light. But, because of the natural finite time duration and the limited frequency band width of a signal or of any particle (no matter whether photon, electron, atom or molecule) the construction of a time machine which would manipulate the past is not possible. Cause and effect cannot be interchanged. Even so, the fact that there are verifiable spaces like mountain tunnels, in which no time exists although they can be traveled through, remains thrilling.

Bibliography

1 Max Born, Einstein's Theory of Relativity, Dover Publications (1962).

2 M. Fayngold, Special Relativity and Motions Faster than Light, Wiley-VCH, Weinheim, 219–223 (2002).

3 Aurelius Augustinus, Confessiones.

4 A. Enders and G. Nimtz, J. Phys. I, France, **2**, 1693 (1992).

5 A. Ranfagni et al., Appl. Phys. Lett. **58**, 774 (1991).

6 R. U. Sexl and H. K. Urbantke, Relativity, Groups, Particles, Springer, Wien, New York (2001).

7 G. Nimtz, Prog. Quantum Electron., **27**, 417 (2003).

8 A. Ranfagni et al., Phys. Rev. E **48**, 1453 (1994).

9 A. Steinberg, P. Kwiat, R. Chiao, Phys. Rev. Lett. **68**, 2421 (1992); A. Steinberg, P. Kwiat, R. Chiao, Phys. Rev. Lett. **71**, 708 (1993).

10 M. Büttiker and S. Washburn, Optics: Ado about nothing much?, Nature, **422**, 271–272 (2003).

11 S. Collins, D. Lowe and J. R. Barker, J. Phys. C **20**, 6213 (1987).

12 G. Nimtz and A. Haibel, Ann. Phys. (Leipzig), **11**, 163 (2002).

13 L. Brillouin, Wave Propagation and Group Velocity, Academic Press, New York and London (1960), p. 79.

14 I. Newton, Philosophia Naturalis Principia Mathematica, (1687), German edition: J. Ph. Wolfers, Berlin, p. 25 (1872).

15 G. Nimtz, in Lecture Notes in Physics, Special Relativity, J. Ehlers and C. Laemmerzahl (Eds.), Springer Berlin, New York (2006), **702**, 506–534.

16 M. Alonso and E. J. Finn, Fundamental University Physics, Vol. III, Addison-Wesley, Reading MA (1969).

17 Th. Hartman, J. Appl. Phys. **33**, 3427 (1962).

18 C. K. Carniglia and L. Mandel, Phys. Rev. D **3** (1971).

19 S. T. Ali, Phys. Rev. D **7** 1668 (1973).

20 R. P. Feynman, QED The Strange Theory of Light and Matter, Princeton University Press, Princeton USA (1985).

21 A. A. Stahlhofen and G. Nimtz, Europhys. Lett. **76**, 189 (2006).

22 A. Haibel, G. Nimtz, A. A. Stahlhofen, Phys. Rev. E **63**, 047601 (2001).

23 F. Goos and H. Hänchen, Ein neuer fundamentaler Versuch zur Totalreflexion, Ann. Phys. (Leipzig) (6) **1**, 333 (1947);

Zero Time Space: How Quantum Tunneling Broke the Light Speed Barrier. G. Nimtz and A. Haibel

ISBN: 978-3-527-40735-4

F. Goos and H. Lindberg–Hänchen, Neumessung des Strahlversetzungseffektes bei Totalreflexion, Ann. Phys. (Leipzig) (6) **5**, 251 (1949).

24 D. Müller, D. Tharanga, A.A. Stahlhofen and G. Nimtz, Europhysics Letters **73**, 526 (2006).

25 Sommerfeld, A., Lectures on Theoretical Physics, Optics. Academic Press Inc. US (1954).

26 D. T. Emerson, National Radio Astronomy Observatory, The Work of Jagadis Chandra Bose, 100 Years of mm-Wave Research (1998) (Photograph from Acharya Jagadis Chandra Bose, Birth Centenary, 1858–1958. Calcutta: published by the Birth Centenary Committee, printed by P. C. Ray, November 1958.)

27 F. de Fornel, Evanescent Waves: from Newtonian Optics to Atomic Optics, Springer Verlag, Berlin/Heidelberg/New York (2001).

28 Proceedings of the 22nd Solvay Conference in Physics The Physics of Communication, European Cultural Center of Delphi, Delphi, 24–29 November 2001; IEEE J. Sel. Top. Quantum Electron., **9**, (1) 79 (2003).

29 Th. Martin and R. Landauer, Phys. Rev. A **45**, 2611 (1992).

30 S. Longhi et al., Phys. Rev. E **64**, 055 602 (2001).

31 F. E. Low and P. F. Mende, Ann. Phys. **210**, 380 (1991).

32 A. Zeilinger, Nature, **401**, p. 680 (1999).

33 P. Mittelstaedt, Philosophical Problems of Modern Physics, D. Reidel, Dordrecht, Holland 1976.

34 A. Haibel and G. Nimtz, Ann. Phys. (Leipzig) **10** (2001) 8, 707–712.

35 C. R. Leavens and G. C. Aers, Phys. Rev. B **40**, 5387 (1989).

36 W. M. Robertson, J. Ash, and J. M. McGaugh, Am. J. Phys. **70** p. 687 (2002).

37 S. Yang, J. H. Page, Z. Liu, M. L. Cowan, C. T. Chan, and P. Sheng, Phys. Rev. Lett. **88**, p. 104301-1 (2002).

38 L. H. Ford und T. A. Roman, Sci. Am., p. 30, January 2000.

Index

Zero Time Space: How Quantum Tunneling Broke the Light Speed Barrier. G. Nimtz and A. Haibel

ISBN: 978-3-527-40735-4